Reference Sources in Science and Technology

by

Earl James Lasworth

The Scarecrow Press, Inc.,
Metuchen, N.J. 1972

Library of Congress Cataloging in Publication Data

Lasworth, Earl James.
Reference sources in science and technology.

1. Science--Bibliography. 2. Technology--Bibliography. 3. Reference books--Bibliography. I. Title.
Z7401.L36 016.5 72-3955
ISBN 0-8108-0500-6

CONTENTS

INTRODUCTION

The Library Search

Every library user from the high school student to the graduate faculty will be involved at one time or another in a library search, either searching for one small fact or perhaps compiling an extensive subject bibliography.

The process of choosing a research topic and the library search for information about the topic are interrelated activities. As information about the topic is uncovered, the research topic may be redefined, and new reference books and periodicals will have to be read and evaluated for relevance. Unfortunately, students may find so much material on their topic that they have difficulty deciding where to begin reading, or the unavailability of one magazine article may discourage them from looking elsewhere for the same information.

A library search seems most difficult before defining the research topic. In order to begin the search, the searcher has to ask himself several questions: "Do I know enough about the subject I am searching?"; What limitations should be placed on the search?"; "Is the material found in the search needed now or should hard to acquire material be included?"; "Are the citations annotated?"; "What form should the bibliography be in?"; etc.

Reading the library's directory, manual or orientation pamphlets will give some idea of where the more important general reference books and indexes are located. Academic reference librarians are charged especially with the task of helping students make use of the library's collections of books, periodicals and other library materials. In most academic libraries reference librarians have questions of many kinds brought to them by students and these are answered insofar as the resources of the library permit.

Library reference service is often spoken of as if it consisted only of the actual use of reference books in answering questions for students. In its widest and proper context, however, the reference librarian's work includes everything necessary to help the student in his inquiries: the selection of an adequate and suitable collection of reference books; the arrangement and maintenance of the collection in such a way that it can be used easily and conveniently; the making of such files, indexes and clipping collections as are needed to supplement the main card catalog and the book collection; the training of a capable staff and their supervision to insure skillful and pleasant service; the provision of posted signs, printed directions, lists and bulletins; expert aid in the use of the card catalogs and other records; suggestions as to books to be used for special purposes; instruction of individuals, groups, or classes in the use of reference books and reference methods; and consultation in helping individual students to find some elusive fact, or in editing some inadequate method of library searching. While a large part of this work is advisory, with the purpose of helping the student to help himself, there will always be included, also, much actual research work in looking up questions, particularly those which students, even with some advice and assistance, have found too difficult.

Reference service is not limited to work within the library. Through interlibrary loan, knowledge of outside specialists--individuals, institutions or learned societies, government or public service bureaus, etc.--from whom help can be obtained, can often open up many profitable sources of information to students.

In beginning a search through any library's reference collection, one should always progress from the general to the specific reference book, making a list of relevant subject headings and terms that accurately define the limits of the topic to be searched. In the course of a manual library search, progress can be reviewed continuously; information discovered in the beginning may change the direction to be taken and suggest narrowing or broadening the subject headings to be searched. In an academic library, the searcher is able to locate and choose the best guides to information from a wide range of reference books and indexes covering both current and older publications in science and related subjects. Furthermore, when using printed guides, a searcher can make evaluations of bibliographic citations as they are uncovered by considering the titles of books and articles, authorship, date of publication and the quality of the periodicals listed.

LIBRARY SEARCH FLOW CHART

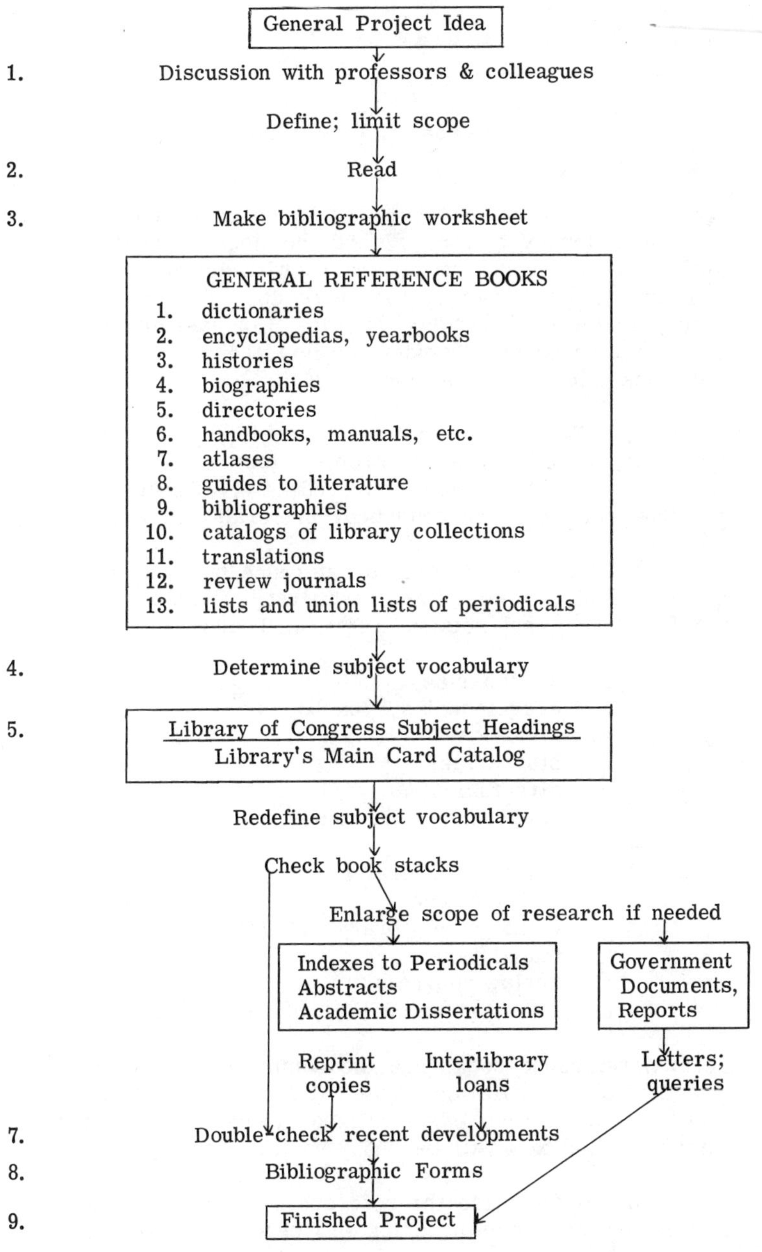

Library Search Flow Chart Checklist

In the preceding chart the boxes represent reference books and library materials, and the lines connecting the boxes show the route of the search with appropriate instructions in the manner of system analysis. The library's role is to help the student help himself. Considerable time and effort will be saved if students follow these steps each time they seek information on any subject.

1. Obtain from instructor a complete, accurate statement of library search problem including language limitations, period of time to be covered, etc. Define and establish the problem to be searched, preferably in writing.

2. Outline the various reference books to be examined: list the most important first. Choose and read general and subject related reference books:

a. dictionaries
b. encyclopedias, yearbooks
c. histories
d. biographies
e. directories
f. handbooks, field guides, guides, manuals, tables
g. atlases
h. guides to literature
i. bibliographies
j. catalogs of library collections
k. translations
l. review journals
m. lists and union lists of periodicals

General reference books use the commonest or most generally accepted scientific and technical English words in their subject headings; more specialized cyclopedias, manuals, guides and indexes will be required for further searching.

3. Systematize the references as they are found. List references, making brief entries on work sheet.

4. Become familiar with the subject and its vocabulary through critical reading of general encyclopedia articles and use of dictionaries or other general reference books.

5. Determine and set up a list of key words, descriptive phrases and specific subject headings related to your subject and revise as the library search progresses. Consult Library of Congress Subject Headings before searching in the library's main card catalog.

6. Enlarge the scope of the search as necessary by adding new periodical titles and document numbers indicated by the clues that appear as the library search progresses in indexing and abstracting publications.

a. Start with most recent index issues available.

b. Remember that subject headings in indexing and abstracting publications may change from year to year within a set of indexes. Consider time lag between publication date of periodical articles and indexing date for abstracts and indexes searched.

c. Examine copies of periodicals too recent to be covered by indexes.

7. Examine the most recent periodical survey or review articles about the subject for bibliography as a source list of additional specialized nomenclature and current information.

8. Review references and transcribe bibliographical citations according to standards appropriate for subject searched.

9. Assemble the references and type the report.

10. At any time during the library search, discuss with professors and colleagues any problems encountered regarding any book or library material.

In addition to the "how to" instructions for library searching, other factors should be mentioned. Nathan Grier Parke, author of Guide to the Literature of Mathematics and Physics Including Related Works on Engineering Science (2nd rev. ed., New York, Dover, 1958), suggests "there are eight important traits that a library searcher should possess before attempting a systematic search."

1) Imagination. This insures that all places have been investigated for the needed information.

2) Mental flexibility. This allows for rapid adjustment to new ideas and possibilities.

3) Thoroughness is needed so that the probability of passing up useful information can be reduced.

4) Orderliness. The search can be a success only if records are kept of exactly what has been searched, what has been found, and where the information was found.

5) Persistency. This eliminates the possibility of giving up too soon. It is useless without the first trait, imagination.

6) A keen sense of observation in order to help turn up unsuspected clues in the search. With the rapid change of scientific terms, this trait is very important.

7) Judgment. This keeps the search progressing along the right path rather than on a tangent.

8) Accuracy. A lack of accuracy in citations and quotations can cause hours of needless work and in many cases result in a useless bibliography.

Chapter I

PURPOSE OF SEARCH

The purpose of a literature search is to bring together all the published information that can be found on a given subject. The search will produce a bibliography that documents the "state-of-the-art" as of a particular date.

To use the complete facilities of any library effectively requires skill which can be learned by any student. In order to develop familiarity and experience in locating facts and information in a library, a literature search project on a topic of interest in the general field of new developments in science and technology is recommended. Individual choice of subject should be based on the availability of reference books such as dictionaries, encyclopedias, handbooks, manuals, guides, atlases, periodicals, government reports and documents and other publications. As long as the information from any published source is the most recent, accurate and factual, it should be included in the "state-of-the-art" bibliography.

Access to the book collection can be gained through subject headings in the library's card catalog. Access to any periodical article is gained through the various periodical indexes and abstracts. Before the card catalog and indexes are consulted, there are several factors to be considered in order to decide where to begin.

In setting up the plan of procedure for a search, the first step is to define what is to be sought. Before any start is made, a class session should be devoted to reaching a mutual understanding as to the scope of the investigation, points to be emphasized, and the possibility of related material that might be pertinent. The known history of a scientific or technical development may be sufficiently definite to make it safe to establish a limiting historical date beyond which it would be unrewarding to go. Once the subject is defined for each student, the outline for the search can be developed.

Define the problem to be searched in the library, preferably in writing.

Analyze the problem. This means establishing the search purpose, bounding the subject area, and setting limits on the time period to be covered. To make an intelligent library search, the searcher must know its purpose as well as the subjects involved. If in reading about the Apollo lunar flight, you find the phrase "automatic landing system" and need to know its meaning, where would you look? The key words and concepts need to be defined, so the next logical step in your search problem is to obtain correct definitions.

Dictionaries

A modern unabridged dictionary is arranged alphabetically just like the general encyclopedia, giving information about the thing as well as the word. A surprising amount of encyclopedic information has been added to dictionaries in the form of special lists and good illustrations. However, instead of covering a wide variety of subjects in depth, the main function of a general dictionary is to define words by showing their historical development, varied meanings and uses.

Desk dictionaries which are edited, revised and updated frequently include many new words and new meanings. Some desk dictionaries are completely new, others are abridged versions of an unabridged dictionary. All give clear, concise definitions, line drawings, antonyms, synonyms, and etymologies authoritatively. The following titles may help you to define and clarify your search problem:

Barnhart, C. L., et al. World book dictionary. Chicago, Field Enterprises Educ. Corp., 1970. 2 v.

Funk & Wagnalls Dictionary Staff, ed. Funk & Wagnalls standard college dictionary. New York, Funk & Wagnalls, 1969.

________. Funk & Wagnalls standard encyclopedic college dictionary. New York, Funk & Wagnalls, 1968.

Funk & Wagnalls Editors. Funk & Wagnalls standard dictionary of the English language. International Ed. New York, Funk & Wagnalls, 1968. 2 v.

Gove, Philip B., ed. Role of the dictionary. New York, Bobbs-Merrill, 1967.

MacLeish, A. and Glorfeld, Louis E. Dictionary and usage: a book of readings. New York, Holt, Rinehart & Winston, 1968.

Merriam Co. Webster's third new international dictionary unabridged. Ed. by Gove, Philip B. et al. Springfield, Mass., Merriam, 1966.

_______. Webster's new world dictionary of the American language. 2nd ed. New York, Popular Library, Inc., 1970.

Pei, Mario, ed. Language of the specialists: communications guide to twenty different fields. New York, Funk & Wagnalls, 1966.

Random House, ed. Random House dictionary of the English language: the college edition. New York, Random House, 1968.

Roget, Peter M. New Roget's thesaurus in dictionary form. Rev. & enl. Ed. by Lewis, Norman. New York, Medallion Books, Berkeley Publishing Corp., 1969.

_______. Original Roget's thesaurus of English words and phrases. Ed. by Dutch, Robert A. New York, Dell Publishing Co., 1966.

A general dictionary may provide sufficient information, but a specialized subject dictionary will probably yield more specific meanings. Words and phrases found in the literature of science offer the reader a special challenge and responsibility for understanding the vocabulary. Subject dictionaries supply, in addition to more specific definitions and shadings of meaning, many special and technical terms not listed in general dictionaries. When approaching a new subject or topic, it is necessary to first gain a general or overall familiarity. This can be accomplished by trying to classify the subject or information sought as to field of knowledge, level of reading difficulty, and degree of specialization. It may require some initial work to arrive at an understanding of the subject area. Subject dictionaries exist for most scientific fields: biology, chemistry, physics, engineering, etc. Many contain encyclopedic information, illustrations, and bibliographical references. Dictionaries are found in the library card catalog under the subject, for example:

Science--Dictionaries; Technology--Dictionaries; Biology--Dictionaries; Chemistry--Dictionaries.

The following science and technology dictionaries are:

1. Those which define the terms in the same language.

2. Bilingual, or those which define the terms in another language, such as Russian-English, or English-German.

3. Polyglot, or those which give the English equivalent of scientific terms in other languages, aiding translation of articles in a foreign language.

GENERAL SCIENCE

Alford, M. H. T. and V. L. Russian-English scientific and technical dictionary. Elmsford, New York, Pergamon, 1970. 2 v.

Ballentyne, Denis William George & Lovett, D. R. A Dictionary of named effects and laws. 3rd ed. New York, Barnes & Noble, 1970.

Bramson, M. ed. Polish-English master dictionary of science and technology. New York, Gordon & Breach, 1969.

Bray, Alexander. Russian-English scientific-technical dictionary. New York, International Univ. Pr., 1969, c1945.

Cagnacci-Schwicker, Angelo. International dictionary of metallurgy, mineralogy, geology, and the mining and oil industries. (Polyglot.) New York, McGraw-Hill, 1970.

DeSola, Ralph. Abbreviations dictionary. Rev. & enl. ed. New York, Hawthorn Books, Inc., 1967.

Dictionary of scientific terms. Alhambra, Calif., Borden (Dennison Pubns.), 1969.

Dorian, Angelo Francis. Dictionary of science and technology. v. 1, English-German. New York, American Elsevier, 1967.

__________. Dictionary of science and technology. v. 2, German-English. New York, American Elsevier, 1970.

Flood, Walter E. & West, Michael. Elementary scientific & technical dictionary. 3rd ed. New York, Fernhill House, 1969.

Graham, Elsie C. Basic dictionary of science. New York, Macmillan, 1967.

Heald, J. H. The Making of test thesaurus of scientific and engineering terms. Springfield, Va., Clearinghouse, 1967. (AD661-001.)

Hogben, Lancelot Thomas. The Vocabulary of science. New York, Stein & Day, 1970, c1969.

Horne, J. Basic vocabulary of scientific & technical German. Elmsford, N.Y., Pergamon, 1970.

Mandel, Siegfried. Dictionary of science. New York, Dell, 1969.

Mullen, William B. Dictionary of scientific word elements. Totowa, N.J., Littlefield, 1969.

Speck, George E., ed. Compact science dictionary. Rev. ed. Greenwich, Conn., Fawcett World (Premier), 1969.

Speck, Gerald E. Dictionary of science terms. New York, Hawthorn Books, 1965.

Thomas, Robert C. and Crowley, Ellen T., eds. Acronyms and initialisms dictionary; a guide to alphabetic designations, contractions and initialisms: associations, aerospace, business, electronic, governmental, international, labor, military, public affairs, scientific, societies, technical, transportation, United Nations. 3rd ed. Detroit, Gale Research Co., 1970.

Tver, David F., ed. Dictionary of business and scientific terms. 2nd ed. Houston, Texas, Gulf Publishing, 1968.

U.S. National Aeronautics and Space Administration. NASA thesaurus; subject terms for indexing scientific and technical information. Washington, D.C., U.S.G.P.O., 1967. 3 v. (NASA SP-7030.)

Zimmerman, Mikhail G. Russian-English translator's dictionary; a guide to scientific and technical usage. New York, Plenum, 1967.

MATHEMATICS

Aleksandrov, A. D., et al, eds. Mathematics: Its content, methods and meaning. 2nd ed. Cambridge, Mass., MIT Press, 1969. 3 v.

Baker, C. C. Dictionary of mathematics. New York, Hart, 1966.

Breuer, Hans. Dictionary for computer languages. Automatic Programming Information Center. New York, Academic Press, 1966. (Studies in Data Processing, v. 6).

Chandor, Anthony. A Dictionary of computers. Baltimore, Md., Penguin, 1970.

Feys, R. & Fitch, F. B., eds. Dictionary of symbols of mathematical logic. New York, Humanities, 1969.

Freund, John E. and Williams, F. J. Dictionary-outline of basic statistics. New York, McGraw-Hill, 1966.

Herland, Leo. Dictionary of mathematical sciences. 2nd ed. Rev. (v. 1, German-English; v. 2, English-German.) New York, Ungar, 1965-66. 2 v.

Hogben, Lancelot. Mathematics for the millions. rev. ed. New York, Norton, 1968.

James, Glenn & Robert C. Mathematics dictionary. (Polyglot.) 3rd ed. New York, Van Nostrand-Reinhold, 1968.

Kotz, Samuel. Russian-English dictionary and reader in the cybernetical sciences. New York, Academic Press, 1966.

__________ & Hoeffding, Wassily. Russian-English dictionary of statistical terms and expressions and Russian reader in statistics. Chapel Hill, N.C., Univ. of N.C. Press, 1964.

Krueger, K. & H. Data processing dictionary. (English, German.) New York, Adler, 1968.

MacIntyre, Sheila & Witte, E. German-English mathematical vocabulary. 2nd ed. New York, Wiley (Interscience), 1966.

Marks, Robert W. New mathematics dictionary and handbook. New York, Grosset & Dunlap, 1964.

Millington, William & Alaric T. Dictionary of mathematics. Cranbury, N. J., A. S. Barnes, 1966.

Paenson, I. Glossary of statistical theory. (English-French-Spanish-Russian.) Elmsford, N. Y., Pergamon, 1970.

Sippl, Charles J. Computer dictionary and handbook. Indianapolis, Ind., Sams, 1966.

Spencer, Donald D. Computer programmer's dictionary and handbook. Waltham, Mass., Ginn, College Division, 1968.

__________. A Thesaurus of computer names. Daytona Beach, Fla., Abacus Computer Corp., c1970.

Trollhann, Lilian & Wittman, A., eds. Dictionary of data processing. (Polyglot.) New York, American Elsevier, 1965.

Weik, Martin H. Standard dictionary of computers and information processing. New York, Hayden Book, 1969.

ASTRONOMY

Chiu, Hong-Yee, ed. Chinese-English and English-Chinese astronomical dictionary. New York, Plenum (Consultants Bureau), 1966.

Gros, E. & Singer, L. Dictionary of astronomy. (Chinese-English-Russian.) New York, Transatlantic, 1969.

Maloney, Terry. Dictionary of astronomy. New York, Fernhill, 1964.

Moore, Patrick. Amateur astronomer's glossary. New York, Norton, 1967.

Motz, Lloyd. Astronomy: A to Z. New York, Grosset & Dunlap, 1964.

Wallenquist, Ake. Dictionary of astronomical terms. New York, Doubleday (AMS Natural History), 1964.

PHYSICS

Clason, W. E. Dictionary of nuclear science & technology. (Polyglot.) New York, American Elsevier, 1970. Russian Supplement, 1, 1970.

DeVries, Louis & Clason, W. E. Dictionary of pure & applied physics. v. 1, German-English. New York, American Elsevier, 1963.

__________. Dictionary of pure & applied physics. v. 2, English-German. New York, American Elsevier, 1964.

Meetham, A. R. Dictionary of physics: multilingual glossary. Vol. 9. Elmsford, N. Y., Pergamon, 1969.

Sube, R. Dictionary of nuclear physics & technology. (English-French-German-Russian.) Elmsford, N. Y., Pergamon, 1969.

Thewlis, J., et al, eds. Encyclopaedic dictionary of physics: general, nuclear, solid state, molecular chemical, metal and vacuum physics, astronomy, geophysics, biophysics, and related subjects. Elmsford, N. Y., Pergamon, 1961-1964. 9 v. Supplements. 4 v. 1966--

CHEMISTRY

Angelé, Hans-Peter, ed. Four-language technical dictionary of chromotography. Elmsford, N. Y., Pergamon, 1971.

Ballentyne, Denis William George. A Dictionary of named effects and laws in chemistry, physics and mathematics. 3rd ed. London, Chapman & Hall, 1970.

Cahn, Robert S. Introduction to chemical nomenclature. 3rd ed. New York, Plenum, 1968.

Cooke, Edward I., ed. Chemical synonyms and trade

names; a dictionary and commercial handbook. 7th ed. Cleveland, Chemical Rubber Co., 1971.

DeVries, Louis. Dictionary of chemistry and chemical engineering. (German/English.) New York, Academic Press, 1970. (Worterbuch der Chemie und der Chemischen Verfahrenstechnik.)

Dictionary of organic compounds. 4th ed. New York, Oxford Univ. Press, 1965. 5 v. Six supplements. 1965-1970.

Ernst, R. Dictionary of chemical terms, v. 1: German-English; v. 2: English-German. New York, Adler, 1969. 2 v.

Flood, Walter E. Dictionary of chemical names. Totowa, N.J., Littlefield, 1967.

Fromherz, Hans and King, Alexander. Chemical terminology. (English-German.) 5th rev. ed. New York, Adler, 1968.

__________. Chemical terminology. (French-English.) New York, Adler, 1968.

Grant, Julius. Hackh's chemical dictionary. 4th ed., rev. New York, McGraw-Hill, 1969.

Hoseh, M. & M. L. Russian-English dictionary of chemistry and chemical technology. New York, Van Nostrand-Reinhold, 1964.

International Union of Pure and Applied Chemistry. Nomenclature of organic chemistry. 2nd ed. New York, Plenum, 1966.

Miall, Lawrence M., ed. Dictionary of chemistry. 4th ed. New York, Wiley (Interscience), 1968.

Neville, Hugh H., et al. New German/English dictionary for chemists. New York, Van Nostrand-Reinhold, 1964.

Powell, Virginia. Chemical formulas and names. New York, Prentice-Hall, 1964.

Rose, Arthur & Elizabeth. Condensed chemical dictionary.

7th ed. New York, Van Nostrand-Reinhold, 1966.

Sobecka, Z., et al, eds. Dictionary of chemistry and chemical technology in six languages. (English, German, Polish, Russian, Spanish, French.) 2nd ed. Elmsford, N.Y., Pergamon, 1966.

Traynham, James. Organic nomenclature: a programmed introduction. New York, Prentice-Hall, 1966.

Wohlauer, Gabriele E. & Gholston, H. D., eds. German chemical abbreviations. New York, Special Lib. Assn., 1966.

EARTH SCIENCES

American Geological Institute. Dictionary of geological terms. New York, Doubleday (Dolphin), 1969.

Challinor, John. Dictionary of geology. 3rd ed. New York, Oxford Univ. Press, 1967.

Knox, Alexander. Glossary of geographical and topographical terms. Detroit, Gale Research Co., 1968.

Lana, Gabriella, et al. Glossary of geographical names. (Polyglot.) New York, American Elsevier, 1967.

Lexicon of geologic names of the United States for 1936-1960. Washington, D.C., U.S. Geological Survey, 1966. 3 v. (USGS Bulletin 1200.)

Merriam Co. Webster's geographical dictionary. Rev. ed. Springfield, Mass., Merriam, 1967.

Monkhouse, Francis J. Dictionary of geography. 2nd ed. Chicago, Aldine, 1970.

Moore, W. G. Dictionary of geography. New York, Praeger, 1967.

Pfannkuch, Hans-Olaf. Dictionary of hydrogeology. In three languages: English, French, German. New York, American Elsevier, 1969.

Runcorn, S. K., ed. International dictionary of geophysics.

Seismology, geomagnetism, aeronomy, oceanography, geodesy, gravity, marine geophysics, meteorology, the Earth as a planet and its evolution. Elmsford, N.Y., Pergamon, 1968. 2 v. and atlas.

Schmieder, Allen, et al. Dictionary of basic geography. Rockleigh, N.J., Allyn and Bacon, 1970.

Stamp, L. Dudley. Glossary of geographical terms. 2nd ed. New York, Wiley, 1966.

__________, ed. Dictionary of geography. New York, Wiley, 1966.

U.S. Naval Oceanographic Office. Glossary of oceanographic terms. 2nd ed. Washington, D.C., 1966. (Special Pub. SP-35.)

U.S. Office of Water Resources Research. Water resources thesaurus. Washington, D.C., U.S.G.P.O., 1966.

Wilmarth, Mary G. Lexicon of geologic names in the United States. St. Clair Shores, Mich., Scholarly, reprint 1968. 2 v.

World Meteorological Organization. International meteorological vocabulary. (English-French-Russian-Spanish.) Geneva, 1966.

BIOLOGICAL SCIENCES

Abercrombie, M., et al. A Dictionary of biology. 5th ed. Baltimore, Md., Penguin, 1966.

Biass-Ducroux, Françoise. Glossary of genetics in English, French, Spanish, Italian, German, Russian. New York, American Elsevier, 1970.

Chamberlin, Willard Joseph. Entomological nomenclature and literature. 3rd ed., rev. & enl. Westport, Conn., Greenwood Press, 1970, c1952.

Chinery, Michael. A Science dictionary of the animal world. New York, Watts, 1969, c1966.

__________. A Science dictionary of the plant world. New York, Watts, 1969, c1966.

Cowan, S. T. Dictionary of microbial taxonomic usage. New York, S-H Service Agency, 1968.

Dumbleton, C. W. Russian-English biological dictionary. New York, Plenum (Consultants Bureau), 1965.

Fulling, Edmund H. Index to botany as recorded in the Botanical Review, 1935-1959. New York, S-H Service Agency, 1967.

King, Robert C. Dictionary of genetics. New York, Oxford Univ. Press, 1968.

Lanjouw, Joseph, et al. International code of botanical nomenclature, adopted by the 10th International Botanical Congress, Edinburgh, 1964. New York, S-H Service Agency, 1966.

Leader, Robert W. & Isabel. Dictionary of comparative pathology and experimental biology. Philadelphia, Saunders, 1971.

Leftwich, A. W. Dictionary of zoology. 2nd ed. New York, Van Nostrand-Reinhold, 1967.

McVaugh, Rogers, et al. Annotated glossary of botanical nomenclature. New York. S-H Service Agency, 1968.

National Research Council. Committee on Animal Nutrition. Biological energy interrelationships and glossary of energy terms. Washington, D. C., National Academy of Science, 1966.

Pennak, Robert Wilson. Collegiate dictionary of zoology. New York, Ronald Press, 1964.

Peters, James A. Dictionary of herpetology. New York, Hafner, 1964.

Snell, Walter Henry. A Glossary of mycology. Cambridge, Harvard Univ. Press, 1971.

Stearn, William T. Botanical Latin: history, grammar, syntax, terminology, and vocabulary. New York, Hafner, 1966.

Thomson, Arthur Landsborough, ed. New dictionary of birds. New York, McGraw-Hill, 1964.

Uphof, J. C. Dictionary of economic plants. 2nd ed. New York, S-H Service Agency, 1968.

Usher, George. Dictionary of botany. New York, Van Nostrand-Reinhold, 1966.

Willis, John C. Dictionary of the flowering plants & ferns. Ed. by Shaw, H. K. A., 7th ed. New York, Cambridge Univ. Press, 1967.

MEDICAL SCIENCES

Blakiston's new Gould medical dictionary. Editor-in-Chief, Arthur Osol. 3rd ed. New York, McGraw-Hill, 1971.

Bolander, Donald O. & Bisdorf, Rita, comp. Instant spelling medical dictionary. Mundelein, Ill., Career Inst., 1970.

Cape, Barbara F., ed. Baillière's nurses dictionary. 17th ed. Baltimore, Maryland, Williams & Wilkins, 1968.

Cole, Frank. The Doctor's shorthand. Philadelphia, Saunders, 1970.

Current procedural terminology. Ed. by Gordon, Burgess L. 2nd ed. Chicago, American Medical Association, 1970.

Dorland, William A. Dorland's illustrated medical dictionary. 24th ed. Philadelphia, Saunders, 1965.

__________. Dorland's pocket medical dictionary. 21st ed. Philadelphia, Saunders, 1968.

Duncan, Helen A. Duncan's dictionary for nurses. New York, Springer Publishing Co., Inc., 1971.

Etter, Lewis E. Glossary of words and phrases used in radiology, nuclear medicine and ultrasound. 2nd ed. Springfield, Ill., C. C. Thomas, 1970.

Fitch, Grace E. and Dubiny, Mary Jane, eds. Macmillan

dictionary for practical and vocational nurses. New York, Macmillan, 1966.

Leider, Morris & Rosenblum, M. Dictionary of dermatological words, terms and phrases. New York, McGraw-Hill, 1968.

Lejeune, Fritz and Bunjes, W. E. Dictionary for physicians. v. 1, German-English; v. 2, English-German. 2nd ed. New York, Grune, 1968-1969.

Mueller, Peter B. German-English; English-German: a dictionary of professional terminology of speech pathology & audiology. Springfield, Illinois, C. C. Thomas, 1967.

Oakes, Lois. Livingstone's pocket medical dictionary. 10th rev. ed. New York, Landau, 1966.

Reissner, Albert & Wade, Carlson. Dictionary of sexual terms. Bridgeport, Conn., Associated Booksellers, 1968.

Rothenberg, Robert E. The New American medical dictionary & health manual. rev. ed. New York, World, 1970, c1962.

Ruiz Torres, F. Spanish-English and English-Spanish medical dictionary. 3rd ed. New York, Heinman, 1965.

Schmidt, Jacob E. Paramedical dictionary: a practical dictionary for the semi-medical and ancillary medical professions. Springfield, Ill., C. C. Thomas, 1969.

__________. Police medical dictionary. Springfield, Ill., C. C. Thomas, 1968.

__________. Practical nurses medical dictionary. Springfield, Ill., C. C. Thomas, 1968.

Skinner, Henry Alan. The Origin of medical terms. 2nd ed. New York, Hafner, 1970, c1961.

Sliosberg, A. Dictionary of pharmaceutical science and techniques. v. 1, Industrial techniques; v. 2, Materia medica. New York, American Elsevier, 1968. 2 v.

__________. Elsevier's medical dictionary in five languages. (English-American-French-Italian-Spanish-German.) New

York, American Elsevier, 1963.

Stedman, Thomas L., et al. Stedman's medical dictionary. 21st ed. Baltimore, Maryland, Williams & Wilkins, 1966.

Steen, Edwin Benzel. Medical abbreviations. 3rd ed. Philadelphia, F. A. Davis, 1971, c1960.

Strand, Helen R. An Illustrated guide to medical terminology. Baltimore, Maryland, Williams & Wilkins, 1968.

Taber, Clarence W. Cyclopedic medical dictionary. 11th ed. Philadelphia, F. A. Davis, 1969.

Unseld, D. W. German-English & English-German medical dictionary. 5th ed. New York, Heinman, 1968.

Veillon, E. and Nobel, A., eds. Medical dictionary. (English, German, French.) 5th rev. ed. New York, Springer-Verlag, 1969.

Winek, Charles L., et al. 1971 Drug abuse reference; general terms, drug abuse terms, user slang, drugs, chemicals, education source guide. Bridgeville, Pa., Bek Technical Pubns., 1971.

AGRICULTURAL SCIENCES

Boerhave Beekman, W. Elsevier's wood dictionary in seven languages: American, French, Spanish, Italian, Swedish, Dutch, and German. New York, American Elsevier, v. 1, 1964; v. 2, 1966; v. 3, 1968.

Farrall, Arthur W. and Albrecht, Carl F. Agricultural engineering: a dictionary and handbook. Danville, Ill., Interstate, 1965.

Merino-Rodriquez, Manuel. Lexicon of parasites and diseases of livestock. (Polyglot.) New York, American Elsevier, 1964.

__________, ed. Lexicon of plant pests and diseases. New York, American Elsevier, 1966.

Muller, Conrad A., comp. Glossary of sugar technology. (English, French, Spanish, Swedish, Dutch, German,

Italian, Danish.) New York, American Elsevier, 1970.

Nijdam, J. and Ministry of Agriculture and Fisheries. Elsevier's dictionary of horticulture in nine languages. (English-French-Dutch-German-Danish-Swedish-Spanish-Italian-Latin.) New York, American Elsevier, 1970.

U.S. National Agricultural Library. Agricultural-biological vocabulary. Washington, D.C., 1967. 2 v. Supplement, 1968--

Ussovsky, B. N. and Linnard, W. Comprehensive Russian-English agricultural dictionary. 2nd rev. ed. Elmsford, N.Y., Pergamon, 1967.

Weck, J. Dictionary of forestry. New York, American Elsevier, 1966.

ENGINEERING SCIENCES

Allied Radio Corp. Technical staff. Dictionary of electronic terms. Ed. by Beam, Robert E. Chicago, Allied Radio, 1968.

Ansteinsson, J. and Andreasson, A. T. Norwegian-English, English-Norweigian technical dictionary. New York, Heinman, 1964-1966. 2 v.

Bender, Arnold E. Dictionary of nutrition & food technology. 3rd ed. Hamden, Conn., Shoe String Press, (Butterworth), 1965.

Bindmann, Werner, ed. Dictionary of semiconductor physics & electronics. (German-English, English-German.) Elmsford, N.Y., Pergamon, 1966.

Booth, Kenneth McIvor. Dictionary of refrigeration and air conditioning. New York, American Elsevier, 1970.

Buksch, H. Dictionary of civil engineering and construction machinery and equipment. v. 1: German-English; v. 2: English-German. New York, Adler, 1967-1969. 2 v.

Burger, Erich. Technical dictionary of data processing, computers, office machines. Elmsford, N.Y., Pergamon, 1970.

Carter, Harley. Dictionary of electronics. New York, Hart, 1967.

Clason, W. E. Dictionary of chemical engineering. v. 1: chemical engineering & laboratory equipment; v. 2: chemical engineering, processes & products. New York, American Elsevier, 1969. 2 v.

__________. Dictionary of electronics & waveguides. (Polyglot.) 2nd ed. New York, American Elsevier, 1966.

__________. Dictionary of metallurgy. (Polyglot.) New York, American Elsevier, 1967.

__________. Electrotechnical dictionary. (Polyglot.) New York, American Elsevier, 1965.

__________. Lexicon of international & national units. New York, American Elsevier, 1964.

Craeybeckx, A. S. H. Dictionary of photography. (Polyglot.) New York, American Elsevier, 1965.

Crispin, Frederic S. Dictionary of technical terms. 11th ed. rev. New York, Bruce Pub., 1970.

Cusset, Francis. English-French and French-English technical dictionary. Rev. ed. New York, Chemical Pub. Co., 1967.

Darcy, Harry L. German-English aerospace dictionary. New York, Adler, 1965.

__________, et al. Russian-English aerospace dictionary. New York, Heinman, 1965.

Denti, Renzo. Italian-English, English-Italian technical dictionary. 6th ed. New York, W. S. Heinman, 1970.

DeVries, Louis. English-German technical & engineering dictionary. 2nd ed. New York, McGraw-Hill, 1968.

Dorian, A. F. Dictionary of aeronautics. (Polyglot.) New York, American Elsevier, 1964.

__________. Dictionary of industrial chemistry. (Polyglot.) New York, American Elsevier, 1964. 2 v.

Engstroem, E. Swedish-English, English-Swedish technical dictionary. New York, Heinman, 1964-1967. 2 v.

Ernst, Richard. German-English & English-German dictionary of industrial technics. New York, Heinman, 1964-1965. 2 v.

Fairchild's dictionary of textiles. 2nd ed. New York, Fairchild Pubn., 1967.

Gerrish, Howard. Technical dictionary. Homewood, Ill., Goodheart, 1968.

__________. Electricity, electronics dictionary. South Holland, Ill., Goodheart-Willcox, 1970.

Gilpin, Alan. Concise encyclopaedic dictionary of fuel technology. New York, American Elsevier, 1970.

__________. Dictionary of fuel technology. New York, Philosophical Lib., 1970.

Gleicher, Norman & Funk & Wagnalls Dictionary Staff, eds. Dictionary of electronics. New York, Funk & Wagnalls, 1969.

Glossary: water and waste water control engineering. New York, American Public Health Assn., 1969.

Graf, Rudolf E. Modern dictionary of electronics. 3rd ed. Indianapolis, Ind., Sams, 1968.

Groves, H. W. Aeronautical technical dictionary: French-English, English-French. New York, Fernhill, 1966.

Hawley, Gessner F. & Alice, W. Hawley's technical speller. New York, Van Nostrand-Reinhold, 1964.

Hohn, Eduard. Dictionary of electrotechnology. (German and English.) New York, Barnes & Noble (Chapman & Hall), 1966.

Horner, J. G. Dictionary of mechanical engineering terms. 9th rev. ed., New York, Heinman, 1967.

Horten, Hans Ernest. Woodworking in four languages: English-German-French-Spanish. New York, Hart, 1969, c1968.

Hughes, Leslie Ernest Charles, Stephens, R. W. B. & Brown, L. D. Dictionary of electronics and nucleonics. New York, Barnes & Noble, 1970, c1969.

Hyman, Charles J. German-English, English-German astronautics dictionary. New York, Plenum (Consultants Bureau), 1968.

__________. German-English, English-German electronics dictionary. New York, Plenum (Consultants Bureau), 1965.

International Academy of Astronautics. Astronautical multilingual dictionary of the International Academy of Astronautics. New York, American Elsevier, 1970.

Jennings, Ralph E., ed. & comp. The Automotive dictionary. New York, William Dogan Annl. Pub. Asso., 1969.

Konarski, M. M. Russian-English space technology dictionary. Elmsford, N.Y., Pergamon, 1970.

Lendeke, Wolfgang. Four-language technical dictionary of heating, ventilation and sanitary engineering. Elmsford, N.Y., Pergamon, 1971.

Macura, Paul. Russian-English dictionary of electrotechnology and allied sciences. New York, Wiley (Interscience), 1971.

Marks, Robert W., ed. The New dictionary and handbook of aerospace; with special sections on the moon and lunar flight. New York, Praeger, 1969.

Markus, John. Electronics and nucleonics dictionary. 3rd ed. New York, McGraw-Hill, 1967.

Marolli, Giorgio. Italian-English, English-Italian technical dictionary. 9th ed. New York, Heinman, 1968.

Moltzer, J., ed. Elsevier's oilfield dictionary. (Polyglot.) New York, American Elsevier, 1965.

Moritz, Heinrich & Török, Tibor. Technical dictionary of spectroscopy and spectral analysis. Elmsford, N.Y., Pergamon, 1971.

Moser, Reta C. Space-age acronyms; abbreviations and designations. 2nd ed. New York, Plenum Press, 1970.

Muller, W., ed. Technical dictionary of automotive engineering. (Polyglot.) Elmsford, N.Y., Pergamon, 1964.

Nayler, J. L. Dictionary of astronautics. Totowa, N.J., Littlefield, 1964.

__________ & G. H. Dictionary of mechanical engineering. New York, Hart, 1967.

Neidhardt, Peter. Four language dictionary of electronics in English, German, French, Russian. Elmsford, N.Y., Pergamon, 1967.

__________. Television engineering & television electronics technical dictionary. (Polyglot.) Elmsford, N.Y., Pergamon, 1964.

Piraux, Henry. French-English, English-French dictionary of electrotechnic, electronics and allied fields. New York, Heinman, 1968. 2 v.

Polanyi, Magda. Technical & trade dictionary of textile terms. (German-English.) 2nd ed. Elmsford, N.Y., Pergamon, 1967.

Pugh, E. Dictionary of acronyms and abbreviations in management, technology, information science. 2nd ed. rev. Hamden, Conn., Shoe String (Archon), 1970.

Richter, Gunter, ed. Dictionary of optics, photography & photogrammetry. (German-English and English-German.) New York, American Elsevier, 1966.

Rodgers, Harold A. Funk & Wagnalls dictionary of data processing terms. New York, Funk & Wagnalls, 1970.

Santholzer, Robert W. Five language dictionary of surface coatings, plating, products finishing, corrosion, plastics, and rubber. (Polyglot.) Elmsford, N.Y., Pergamon, 1969.

Scharf, B. Engineering and its language. London, British Book Centre, 1971.

Scott, John S. Dictionary of civil engineering. rev. ed. Baltimore, Md., Penguin (Pelican), 1967.

Segditsas, P. E. Elsevier's nautical dictionary. (Polyglot.) New York, American Elsevier, 1965-1966. 3 v.

Simons, Eric N. Dictionary of ferrous metals. London, British Book Centre, 1971.

Singer, Lothar, comp. Hydroelectrical dictionary. (Russian-English-German-French.) Scientific Information Consultants (dist. by Transatlantic Arts), 1968, c1967.

Steindler, R. A. The Firearms dictionary. Harrisburg, Pa., Stackpole, 1970.

Stekhoven, G. S. and Valk, W. B. Elsevier's dictionary of metal cutting tools. In seven languages: English/American, German, Dutch, French, Spanish, Italian, Russian. New York, American Elsevier, 1970.

Stenius, Ake. Nomenclature and definitions applicable to radiometric and photometric characteristics of matter. Philadelphia, American Society for Testing and Materials, 1970.

Visser, A. Dictionary of soil mechanics. (Polyglot.) New York, American Elsevier, 1965.

Walther, Rudolf. Multilingual dictionary of mechanics, strength of materials & materials. Elmsford, N.Y., Pergamon, 1969.

__________, ed. Polytechnical dictionary: v. 1, English-German; v. 2, German-English. Elmsford, N.Y., Pergamon, 1968. 2 v.

Weber, Fritz W. Dictionary of high vacuum science & technology. New York, American Elsevier, 1968.

Wernicke, H. Dictionary of electronics, communications & electrical engineering. (English-German; German-English.) New York, Adler, 1969. 2 v.

In defining and analyzing the problem, you must establish the scope of the search paper. Is your topic sufficiently narrowed to allow for adequate treatment, but large enough to find sufficient material? At what point do the instructor's directions, the library material available, and the time for completion meet? The precise subject must be determined not only to prevent needless work, but also to develop a completely relevant bibliography and/or report. It is the student's responsibility to pick a subject about which the library has sufficient information. Usually this can only be done after determining the library's resources. For example, Space flight (TL790) is too large a subject for a term paper, and automatic landing systems is not listed as a subject heading by the Library of Congress and may be too limited. Fortunately, the subject heading Space flight to the moon (TL799.M6) and Project Apollo are listed in Library of Congress Subject Headings. This means any book with the call number TL799.M6 is about the flight to the moon and might contain information on "automatic landing systems."

Chapter II

GENERAL REFERENCE

Outline the various reference books and indexes to be examined: list the most important first.

Included in the science reference book collection are dictionaries, glossaries, encyclopedias, histories, directories, biographies, handbooks, field guides, guides, manuals, tables, atlases, and bibliographies. These reference books are normally consulted for specific information rather than continuous reading.

However, before you look for information in any reference book, go through this reference book checklist:

1. Examine title page for:
   a. subject headings included in title.
   b. authors name and list of degrees, position and titles of earlier books.
   c. publisher.

2. Check date of publication against copyright dates and date of preface or introduction. Usually the earliest date listed indicates when information was compiled.

3. Read preface and/or introduction to obtain author's reasons and objectives for the book including subjects covered, special lists and features, and limitations.

4. Read table of contents to determine how information is organized.

5. Read index, noting cross references to authors, titles and subjects included.

6. Read an article about a subject familiar to you

for accuracy of fact and bibliography.

7. Read several articles about unfamiliar subjects, and compare with similar information in other reference books.

Encyclopedias and Yearbooks

If a particular concept is central to your library search, general encyclopedias and yearbooks may be more useful than a dictionary. Encyclopedia articles are edited to supply introductory and review information on a wide variety of subjects. The yearbook is an annual up-dating and compilation of general and specialized information. Because encyclopedias and yearbooks are limited to knowledge and information as of a specific date, always try to obtain the most recently published. However, older editions should be consulted for information on the state-of-the-art at the date when the encyclopedia was written and for biographical data that might be deleted from more recent editions.

First, locate your subject alphabetically in general encyclopedia articles to become familiar with its magnitude and historical development. This will give you an idea of the relevance of the topic. Then, read all the cross-referenced encyclopedia articles for summary information until you become familiar with the words used to describe the prominent features of the subject. Use an unabridged or subject dictionary to define any word new to you. Develop a list of definitions to be used as "see also" references when you later search for subject headings in the library's card catalog. This list of subject headings will also help when you later search subject encyclopedias and periodical indexes for related information.

An encyclopedia article cannot appear under every synonymous subject heading. Therefore, it is entered under the one most people will think of, with cross references to it under the others. "See also" references help by leading to added information. If you do not find cross references, don't assume that the encyclopedia contains nothing on your subject. Look under as many synonymous subject headings as possible.

Encyclopedia articles usually end with a brief bibliography of some of the important books describing the subject in greater detail. The minimum bibliographical information about each book should include: author's name, with initials, title, place, publisher, and date of publication. Sometimes a printed bibliography on your subject is listed which will save you much of the time you would otherwise spend locating related references. These books should be consulted to further define the subject to be searched. Choose those books with the most recent publication date for your own working bibliography. Books published within the last five years may contain a comprehensive bibliography of other books and periodical articles further describing the subject searched. Any book more than five years old can be deleted from the list, unless the search problem is historical.

In addition to general encyclopedias and yearbooks there are many specialized encyclopedias and cyclopedias of special interest to the searcher. The McGraw-Hill Encyclopedia of Science and Technology is a good example of an encyclopedia on a subject or group of related subjects arranged alphabetically. Scientific subject encyclopedias contain specific data, biographical sketches, useful abbreviations, symbols, tables, diagrams, tests and illustrations. In a good cyclopedia, the editors and contributors will be specialists and their special knowledge will be evident in titles that give some idea of their scope and contents, the choice of subjects, length and authority of articles, fullness and selection of bibliographies, etc. The following list is grouped by subject matter.

ENCYCLOPEDIAS, GENERAL

Benet, William R. The Reader's encyclopedia. 2nd ed. New York, T. Y. Crowell, 1965.

Bridgewater, William & Kurtz, Seymour, eds. Columbia encyclopedia. 3rd ed. Irvington-On-Hudson, N.Y., Columbia Univ. Press, 1963.

Collier's Encyclopedia, with bibliography and index. New York, Crowell-Collier Educational Corp., 1971. 24 v.

Collier's yearbook. New York, Crowell-Collier Educational Corp., 1970.

Cowles Book Company, ed. Cowles volume library. (Orig. title: Cowles comprehensive encyclopedia.) New York,

Cowles, 1969.

Cowles Education Corp. Editors. Cowles encyclopedia of nations. New York, Cowles, 1968.

Encyclopaedia Britannica. Revised annually. Chicago, Encyclopaedia Britannica, Inc., 1970. 24 v.

Britannica yearbook of science and the future, 1969. Chicago, Encyclopaedia Britannica, Inc., 1965--

Encyclopedia Americana. New York, Grolier, Inc., 1970. 30 v.

Facts on file, full news reference service. Annual. New York, Facts on File, v. 1-- 1941--

Grolier universal encyclopedia. Ed. by Martin, Lowell A. New York, Grolier, Inc., 1970. 20 v.

Encyclopedia science supplement. New York, Grolier, Inc. 1965-- (An annual supplement to the Grolier Encyclopedia.)

The McKay one-volume international encyclopedia. Ed. by E. M. Horsley. New York, McKay, 1970.

New standard encyclopedia. Chicago, Standard Educ. Corp., 1970. 14 v.

Pears cyclopaedia; a book of reference and background information for everyday. 79th ed. New York, Schocken, 1970-71.

Taylor, J. W. Lincoln library of essential information. Columbus, Ohio, Frontier Press, 1970.

United Published Corporation. Standard International encyclopedia. New York, United Pub., 1970. 2 v.

World book encyclopedia. Chicago, Field Enterprises Educ. Corp., 1970. 20 v.

Science year: the world book science annual. Chicago, Field Enterprises Educ. Corp., 1965--

SCIENCE, GENERAL

Ackener, Joseph, et al. Pocket encyclopedia of physical science. New York, Golden Press, 1968.

Book of popular science. Rev. annually. New York, Grolier, Inc., 1970. 10 v.

McGraw-Hill encyclopedia of science and technology. rev. ed. New York, McGraw-Hill, 1966. 15 v. Yearbook. 1963--

Neurath, Otto, et al, eds. Foundations of the unity of science: international encyclopedia of unified science. Chicago, Univ. of Chicago Press, v. 1, 1955; v. 2, 1970.

> Feigl, Herbert and Morris, Charles. International encyclopedia of unified science: bibliography and index. Chicago, Univ. of Chicago Press, 1970.

Newman, James R., ed. Harper encyclopedia of science. 2nd ed. New York, Harper & Row, 1967. 2 v.

Popular Science Magazine Editors. Encyclopedia of the sciences. New York, Grosset & Dunlap, 1966.

Science News yearbook 1970. Ed. and comp. by Science Service. New York, Scribner, 1970.

Van Nostrand's scientific encyclopedia. 4th ed. New York, Van Nostrand-Reinhold, 1968.

MATHEMATICS

Jordain, Philip B. Condensed computer encyclopedia. New York, McGraw-Hill, 1969.

Universal encyclopedia of mathematics. New York, Simon & Schuster, 1964.

ASTRONOMY

DeVancouleurs, Gerard H. Reference catalog of bright galaxies. Austin, Texas, Univ. of Texas Pr., 1964.

Hoffleit, D. Catalogue of bright stars. New Haven, Conn., Yale Univ. Observatory, 1964.

Moore, Patrick, ed. Yearbook of astronomy, 1970. rev. New York, Norton, 1970.

Muller, Paul. Concise encyclopedia of astronomy. Chicago, Follett, 1968.

Smithsonian Institution. Astrophysical Observatory. Star catalog. Washington, D. C., U. S. G. P. O., 1966. 4 v.

Weigert, A. and Zimmermann, H. A Concise encyclopedia of astronomy. New York, American Elsevier, 1968.

PHYSICS

Besancon, Robert M. Encyclopedia of physics. New York, Van Nostrand-Reinhold, 1966.

Musset, Paul and Lloret, Antonio. Concise encyclopedia of the atom. Chicago, Follett, 1968.

CHEMISTRY

Carswell, D. J. Introduction to nuclear chemistry. New York, American Elsevier, 1967.

Clark, George L. & Hewley, Gessner G., eds. Encyclopedia of chemistry. 2nd ed. New York, Van Nostrand-Reinhold, 1966.

Encyclopaedia of cybernetics, tr. from the German Lexicon der Kybernetik by Gilbertson, G. New York, Barnes & Noble, 1968.

Hampel, Clifford A. Encyclopedia of electrochemistry. New York, Van Nostrand-Reinhold, 1964.

Kingzett, Charles T. Chemical encyclopedia: a digest of chemistry and its industrial applications. 9th ed. Ed. by Hey, D. H. New York, Van Nostrand-Reinhold, 1967.

Merck index of chemicals and drugs; an encyclopedia for chemists, pharmacists, physicians, and members of allied professions. 8th ed. Rahway, N.J., Merck, 1968.

Organic reactions. New York, Wiley, v. 1-- 1942--

Organic synthesis. New York, Wiley, v. 1-- 1921-- collective volumes. Annual. 1932- published after every tenth annual vol.

Theilheimer, William. Synthetic methods of organic chemistry. New York, Wiley (Interscience), v. 1-- 1946--

Van Nostrand's international encyclopedia of chemical science. New York, Van Nostrand-Reinhold, 1964.

EARTH SCIENCES

Ellis, B. F. & Messina, A. R. Catalogue of foraminifera... New York, American Museum of Natural History, 1940-- (looseleaf). Continued by looseleaf supplements.

Fairbridge, Rhodes Whitmore. Encyclopedia of atmospheric sciences and astrogeology. New York, Van Nostrand-Reinhold, 1967. (Encyclopedia of Earth Sciences Series, v. 2.)

_________. Encyclopedia of oceanography. New York, Van Nostrand-Reinhold, 1966.

_________, ed. The Encyclopedia of geomorphology. New York, Van Nostrand-Reinhold, 1968. (Encyclopedia of Earth Sciences Series, v. 3.)

Firth, Frank E., ed. The Encyclopedia of marine resources. New York, Van Nostrand-Reinhold, 1969.

Gresswell, R. Kay and Huxley, Anthony, eds. Standard encyclopedia of the world's rivers and lakes. New York, Putnam, 1966.

Huxley, Anthony, ed. Standard encyclopedia of the world's mountains. New York, Putnam, 1969.

_________. Standard encyclopedia of the world's oceans and islands. New York, Putnam, 1969.

National Academy of Sciences. IGY World Data Center A:

Oceanography. Catalogue of data in World Data Center A. Washington, D.C., 1957/63. Supplements, 1964-

Sachs, Mosche Y., ed. Worldmark encyclopedia of the nations. 3rd rev. ed. New York, Harper & Row, 1967. 5 v.

Sinkankas, John. Van Nostrand's standard catalog of gems. New York, Van Nostrand-Reinhold, 1968.

Todd, David Keith. The Water encyclopedia. Port Washington, N.Y., Water Information Center, Inc., 1971.

U.S. Bureau of Mines. Minerals yearbook. Washington, D.C. 1862--

U.S. Office of Water Resources Research. Water resources research catalog. Washington, D.C., U.S.G.P.O., 1965--

BIOLOGICAL SCIENCES

Cowles encyclopedia of animals and plants. New York, Cowles, 1968.

Encyclopedia of reptiles & amphibians. New York, TFH Pubns., 1969.

Encyclopedia of seahorses. New York, TFH Pubns., 1969.

Encyclopedia of water plants. New York, TFH Pubns., 1969.

Gray, Peter, ed. Encyclopedia of the biological sciences. 2nd ed., Van Nostrand-Reinhold, 1970.

Hanzak, J. The Pictorial encyclopedia of birds. New York, Crown, 1968.

Hooker, J. D. and Jackson, B. D., eds. Index kewensis: an enumeration of the genera and species of flowering plants from the time of Linnaeus to the year 1885. New York, Oxford Univ. Press. 2 v.

__________. Supplement 1, 1866-1895. Ed. by Durand, T. and Jackson, B. D. 1901-1906.

__________. Supplement 2, 1896-1900. Ed. by Dyer, W. T. 1905.

__________. Supplement 3, 1901-1905. Ed. by Prain, D. 1908.

__________. Supplement 4, 1906-1910. Ed. by Prain, D. 1910.

__________. Supplement 5, 1911-1915. Ed. by Prain, D. 1921.

__________. Supplement 6, 1916-1920. Ed. by Hill, A. W. 1926.

__________. Supplement 7, 1921-1925. Ed. by Hill, A. W. 1929.

__________. Supplement 8, 1926-1930. Ed. by Hill, A. W. 1933.

__________. Supplement 9, 1931-1935. Ed. by Hill, A. W. 1938.

__________. Supplement 10, 1936-1940. Ed. by Hill, A. W. and Salisbury, E. J. 1947.

__________. Supplement 11, 1941-1950. Ed. by Salisbury, E. J. 1953.

__________. Supplement 12, 1951-1955. Ed. by Taylor, G. 1959.

__________. Supplement 13, 1959-1960. Ed. by Taylor, G. 1966.

__________. Supplement 14, 1961-1965. Ed. by Taylor, G. 1970.

Index-catalogue of medical and veterinary zoology. Washington, D.C., U.S.G.P.O., v. 1-18, 1932-52. Supplement, 1953--

Kadans, Joseph M. Modern encyclopedia of herbs. West Nyack, N.Y., Parker, 1970.

Lamb, Edgar & Brian. The Pocket encyclopedia of cacti and succulents in color. New York, Macmillan, 1970, c1969.

Larousse encyclopedia of animal life. New York, McGraw-Hill, 1967.

Menninger, Edwin Arnold. Flowering vines of the world, an encyclopedia of climbing plants. New York, Hearthside, 1970.

Novak, F. A. Pictorial encyclopedia of plants and flowers. New York, International Pubns. Service, 1966.

Shilling, Charles W., ed. Atomic energy encyclopedia in the life sciences. Philadelphia, Saunders, 1964.

Wagner, Robert J. & Abbott, R. Tucker, eds. Van Nostrand's standard catalog of shells. 2nd ed. New York, Van Nostrand-Reinhold, 1967.

Williams, Roger J. & Landsford, Edwin M. Jr., eds. The Encyclopedia of biochemistry. New York, Van Nostrand-Reinhold, 1967.

MEDICAL SCIENCES

Burney, Leroy E., et al, eds. Medical aid encyclopedia for the home. New York, Stravon, 1964.

Family health encyclopedia. Philadelphia, Lippincott, 1970. 2 v.

Hyman, Harold Thomas. The Complete home medical encyclopedia. New York, Hawthorn, 1966.

Parr, John A. and Young, R. A. Parr's concise medical encyclopedia. New York, American Elsevier, 1965.

The Year book of cancer. Chicago, Year Book Medical Pub., v. 1-- 1956/57--

The Year book of dentistry. Chicago, Year Book Medical Pub., v. 1-- 1936--

The Year book of drug therapy. Chicago, Year Book Medical Pub., v. 1-- 1933--

Year book of medicine. Chicago, Year Book Medical Pub., v. 1-- 1901--

Yearbook of pediatrics. Chicago, Year Book Medical Pub., v. 1-- 1933--

AGRICULTURAL SCIENCES

American potato yearbook. Westfield, N.J., C. Stedman Macfarland, Jr., ed. & pub., v. 1-- 1960--

McClure, Robert, ed. Every horse owner's cyclopedia. Huntsville, Texas, I-Tex Pub., 1971.

Miller, William C. and West, Geoffrey P., eds. Encyclopedia of animal care. 8th ed. Baltimore, Md., Williams & Wilkins, 1968.

Nicolaisen, Age. The Pocket encyclopedia of indoor plants in color. New York, Macmillan, 1970, c1969.

Seiden, Rudolph. Livestock health encyclopedia. Ed. by Gough, W. James. 3rd ed. New York, Springer Pubns., 1968.

Seymour, Edward Loomis Davenport, ed. The Wise garden encyclopedia. New York, Grosset & Dunlap, 1970.

Summerhays, Reginald Sheriff. Summerhays' encyclopedia for horsemen. 5th rev. ed. New York, Warne, 1970.

U.S. Department of Agriculture. Yearbook of agriculture. Washington, D.C., 1894--

Veterinarians' blue book. New York, R. H. Donnelley, 1st ed., 1953-- (Formerly Veterinary Drug Encyclopedia and Therapeutic Index, 1953-1966.)

Wyman, Donald. Wyman's gardening encyclopedia. New York, Macmillan, 1971.

Yearbook of veterinary medicine. Chicago, Year Book Medical Pub., v. 1, 1963-1965.

ENGINEERING SCIENCES

Bergaust, Erik, ed. The New illustrated space encyclopedia. New York, Putnam, 1970, c1965.

Cars of the world. Los Angeles, Peterson Pub. Co., 1970-- (title varies)

Casari, Robert B. Encyclopedia of U. S. Military aircraft. Chillicothe, Ohio, Robert Casari, 1970.

Chemical technology: an encyclopedia treatment. The economic application of modern technological developments. New York, Barnes & Noble, 1968. v. 2, 1970.

Cowles encyclopedia of science, industry & technology. Ed. by Cowles Book Company. 2nd ed. New York, Cowles, 1969.

De Galiana, Thomas. Concise encyclopedia of astronautics. Chicago, Follett, 1968.

Encyclopedia of chemical technology. New York, Wiley (Interscience), v. 1-- 1963--

Encyclopedia of instrumentation and control. Editor-in-chief Douglas M. Considine. New York, McGraw-Hill, 1971.

Encyclopedia of polymer science and technology. New York, Wiley (Interscience), v. 1-- 1964--

Focal Press Ltd. Focal encyclopedia of photography. New York, McGraw-Hill, 1969.

Georgano, G. N., ed. The Complete encyclopedia of motorcars; 1885-1968. New York, Dutton, 1968.

Haggerty, James L., ed. Aerospace year book 1970. New York, Spartan, 1970. (title varies; publisher varies)

Jane's all the world's aircraft, 1970-1971. New York, McGraw-Hill, 1970. (title varies)

Jane's fighting ships, 1970-1971. New York, McGraw-Hill, 1970.

Jane's freight containers, 1970-1971. New York, McGraw-Hill, 1970.

Jane's surface skimmer systems, 1970-1971. New York, McGraw-Hill, 1970. (title varies)

Jane's weapon systems, 1970-1971. New York, McGraw-Hill, 1970.

Jane's world railways, 1970-1971. New York, McGraw-Hill, 1970.

Kutateladze, Samson S. and Borishanskii, V. M. Concise encyclopedia of heat transfer. Elmsford, N.Y., Pergamon, 1966.

McGraw-Hill Editors. Encyclopedia of space. New York, McGraw-Hill, 1968.

Mead, William J. Encyclopedia of chemical process equipment. New York, Van Nostrand-Reinhold, 1964.

Merriman, A. D. Concise encyclopaedia of metallurgy. New York, American Elsevier, 1965.

Molloy, E. and Say, M. G., eds. Electrical engineers reference book; a comprehensive work of reference providing a summary of latest practice in all branches of electrical engineering. 11th ed., New York, Transatlantic, 1964.

Motor service's automotive encyclopedia. South Holland, Ill., Goodheart-Willcox, 1970.

Natural and synthetic fibers yearbook. New York, Wiley (Interscience), 1954--

Roes, Nicholas. The Space-flight encyclopedia. Chicago, Follett, 1968.

Simonds, Herbert R., ed. Encyclopedia of plastics equipment. New York, Van Nostrand-Reinhold, 1964.

__________ and Church, James M. Encyclopedia of basic materials for plastics. New York, Van Nostrand-Reinhold, 1967.

Sittig, Marshall, ed. Inorganic chemical and metallurgical process encyclopedia. Park Ridge, N.J., Noyes, 1968.

Smithells, Colin J. Metals reference book. 4th ed. New York, Plenum (IFI), 1967. 3 v.

Snell, Foster D. & Hilton, C. L., eds. Encyclopedia of industrial chemical analysis. New York, Wiley (Interscience), 1966-1970. 10 v.

Ulrey, Harry F. Builders encyclopedia. Indianapolis, Audel, 1970.

Vollmer, Ernst. Encyclopedia of hydraulics, soil & foundation engineering. New York, American Elsevier, 1967.

World car catalogue. Bronxville, N. Y., Herald Books for Automobile Club of Italy, 1970.

Histories

For many historical questions, general encyclopedias, biographical dictionaries, guides to the literature, handbooks, and periodical indexes may be used.

Histories of a subject give the main facts and names in its development, and often contain full bibliographies. General and subject histories of science and technology with comprehensive indexes may be used as encyclopedias of the subject. Searchers interested in the history of science and technology should consult the following appropriate subject areas.

SCIENCE, GENERAL

Asimov, Isaac. Adding a dimension; seventeen essays on the history of science. New York, Doubleday, 1964.

Boas, Marie. The Scientific renaissance, 1450-1630. New York, Harper & Row, 1966.

DeSantillana, Giorgio. The Origins of scientific thought; from Anaximander to Proclus, 600 BC to 500 AD. New York, New American Library, 1970, c1961.

Gilman, William. Science: U. S. A. New York, Viking, 1965.

Hoyt, Edwin P. A Short history of science: v. 1, ancient science; v. 2, modern science. New York, John Day, 1965-66. 2 v.

Merton, Robert King. Science technology & society in seventeenth century England. New York, Harper, 1970.

Nobel Foundation. Prix Nobel en 1966. New York, American Elsevier, 1967.

__________. Prix Nobel en 1967. New York, American Elsevier, 1969.

Pledge, Humphry Thomas. Science since 1500; a short history of mathematics, physics, chemistry, biology. Gloucester, Mass., Peter Smith, 1969, c1959.

Sarton, George Alfred Leon. A History of science. New York, Norton, 1970, c1959. 2 v.

Source books in the history of science series. Cambridge, Mass., Harvard Univ. Pr., 1929-- (early volumes published originally by McGraw-Hill.)

Taton, Rene, ed. History of science. New York, Basic Books, 1963-1966. 4 v. (English translation of his Histoire Générale des Sciences. Presses Universitaires, 1957-64. 3 v. 4.)

MATHEMATICS

Bochner, Salomon. The Role of mathematics in the rise of science. Princeton, N.J., Princeton Univ. Pr., 1966.

Boyer, C. B. A History of mathematics. New York, Wiley, 1968.

Eves, Howard. An Introduction to the history of mathematics. 3rd ed. New York, Holt, Rinehart & Winston, 1969.

Scott, Joseph Frederick. A History of mathematics; from antiquity to the beginning of the nineteenth century. New York, Barnes & Noble, 1969.

Struik, Dirk J. A Concise history of mathematics. 3rd rev. ed. New York, Dover, 1967.

__________, ed. A Source book in mathematics, 1200-1800. Cambridge, Mass., Harvard Univ. Pr., 1969. (Source Books in the History of the Sciences Series.)

ASTRONOMY

Cotter, C. H. A History of nautical astronomy. New York, American Elsevier, 1968.

PHYSICS

Gamow, George. Thirty years that shook physics: the story of quantum theory. New York, Doubleday, 1966.

Gunter, Pete Addison. Bergson and the evolution of physics. Knoxville, Univ. of Tennessee Press, 1969.

Jungk, Robert. Brighter than a thousand suns; a personal history of the atomic scientists. New York, Harcourt Brace, 1970, c1959.

Nobel Foundation. Nobel lectures in physics, 1901-62. v. 1, 1901-1921; v. 2, 1922-1941; v. 3, 1942-1962. New York, American Elsevier, 1964-67. 3 v.

CHEMISTRY

Asimov, Isaac. A Short history of chemistry: an introduction to the ideas and concepts of chemistry. New York, Doubleday, 1965.

Chymia; annual studies in the history of chemistry. Philadelphia, Univ. of Pa. Pr., v. 1-- 1948--

Farber, Eduard. Evolution of chemistry: a history of its ideas, methods, & materials. 2nd ed. New York, Ronald Pr., 1969.

__________, ed. Milestones of modern chemistry: original reports of the discoveries. New York, Basic Books, 1966. (Science & Discovery Series.)

Knight, David M., ed. Classical scientific papers--chemistry. New York, American Elsevier, 1968.

Leicester, Henry M. & Klickstein, Herbert S., eds. Source book in chemistry, nineteen hundred to nineteen fifty. Cambridge, Mass., Harvard Univ. Pr., 1968. (Source Books in the History of the Sciences.)

Lindsay, Jack. The Origins of alchemy in Graeco-Roman Egypt. New York, Barnes & Noble, 1970.

Multhauf, R. P. The Origins of chemistry. Chicago, Univ.

of Chicago Pr., 1968. (Oldburn History of Science Series.)

Nobel Foundation. Nobel lectures in chemistry, 1901-1962. v. 1, 1901-21; v. 2, 1922-41; v. 3, 1942-62. New York, American Elsevier, 1964-67. 3 v.

Partington, James R. History of chemistry. New York, St. Martin, 1962--. 4 v. (v. 1 revised 1970.)

Weeks, Mary E. Discovery of the elements. 7th ed. Ed. by Leicester, Henry M. Easton, Pa., Chemical Education, 1968.

EARTH SCIENCES

Biswas, Asit K. History of hydrology. New York, American Elsevier, 1970.

Hirsch, S. Carl. Mapmakers of America; from the age of discovery to the space era. New York, Viking, 1970.

Idyll, Clarence P., ed. Exploring the ocean world; a history of oceanography. New York, T. Y. Crowell, 1969.

Manning, T. G. Government in science: the U.S. Geological Survey, 1867-1894. Lexington, Ky., Univ. of Ky. Pr., 1967.

Mather, Kirtley F., ed. Source book in geology, nineteen hundred to nineteen fifty. Cambridge, Mass., Harvard Univ. Pr., 1967. (Source Books in the History of Sciences.)

Popkin, Roy. The Environmental Science Service Administration. New York, Praeger, 1967.

Strangway, David W. History of the Earth's magnetic field. New York, McGraw-Hill, 1970.

BIOLOGICAL SCIENCES

Asimov, Isaac. Short history of biology. New York, Doubleday, 1964.

Bradbury, Savile. The Evolution of the microscope. Elmsford, N.Y., Pergamon, 1967.

Dunn, L. C. A Short history of genetics. New York, McGraw-Hill, 1966.

Gardner, E. J. History of biology. 2nd ed., Minneapolis, Burgess, 1965.

Lechevalier, H. A. & Solotorovsky, M. Three centuries of microbiology. New York, McGraw-Hill, 1965.

Oppenheimer, Jane M. Essays in the history of embryology and biology. Cambridge, Mass., M.I.T. Press, 1967.

Ravin, Arnold W. The Evolution of genetics. New York, Academic Press, 1965.

Sturtevant, Alfred H. A History of genetics. New York, Harper & Row, 1965.

MEDICAL SCIENCES

Cumston, Charles Greene. An Introduction to the history of medicine from the time of the Pharaohs to the end of the XVIIIth century... New York, Humanities, 1926, reprinted 1968.

Fishbein, Morris. A History of the American Medical Association, 1847 to 1947. New York, Kraus Reprint, 1969.

Herrlinger, Robert. History of medical illustrations from antiquity to 1600. London, British Book Centre, 1971.

Kett, Joseph F. The Formation of the American medical profession: the role of institutions, 1780-1860. New Haven, Conn., Yale Univ. Pr., 1968. (Studies in the History of Science & Medicine, No. 3.)

Nobel Foundation. Nobel lectures in physiology-medicine, 1901-1962. v. 1, 1901-21; v. 2, 1922-41; v. 3, 1942-62. New York, American Elsevier, 1964-67. 3 v.

Talbott, John Harold. A Biographical history of medicine: excerpts and essays on the men and their work. New York, Grune, 1970.

Wain, Harry. A History of preventive medicine. Springfield, Ill., C. C. Thomas, 1970.

Walsh, James Joseph. Makers of modern medicine. Freeport, N. Y., Books for Libs., 1907, reprinted 1970.

AGRICULTURAL SCIENCES

Carrier, Lyman. The Beginnings of agriculture in America. New York, McGraw-Hill, 1923. Reprint: New York, Johnson Reprint, 1968. (History of American Economy Series.)

Hyams, Edward. A History of gardens and gardening. New York, Praeger, 1971.

Moore, Ernest G. The Agricultural Research Service. New York, Praeger, 1967.

True, Alfred Charles. A History of agricultural education in the United States 1785-1925. New York, Arno Press, 1969.

__________. A History of agricultural experimentation & research in the United States 1607-1925. New York, Johnson Reprint, 1970. (Originally U. S. Dept. of Agriculture, U. S. G. P. O., 1937.)

__________. A History of agricultural extension work in the United States, 1785-1923. New York, Arno Press, 1969.

U. S. Department of Agriculture. A Guide to understanding the United States Department of Agriculture. Washington, D. C., U. S. G. P. O., 1965.

Volin, Lazar. A Century of Russian agriculture; from Alexander II to Khrushchev. Cambridge, Mass., Harvard Univ. Pr., 1970.

ENGINEERING SCIENCES

Beaumont, Anthony. Traction engines and steam vehicles in pictures. New York, A. M. Kelley, 1969.

Bernal, John Desmond. Science and industry in the nineteenth century. Bloomington, Ind., Indiana Univ. Pr., 1970, c1953.

Bordaz, Jacques. Tools of the old and new stone age. Garden City, N.Y., American Museum of Natural History Press, 1970.

Brockmann, Maxwell C. and Elder, Albert L., eds. The History of penicillin production. New York, American Institute of Chemical Engineers, 1970. (Chemical Engineering Progress Symposium Series, no. 100, v. 66, 1970.)

Daumas, Maurice, ed. A History of technology and invention; progress through the ages. New York, Crown, 1970, c1969. 2 v.

Davenport, William H. & Rosenthal, D., eds. Engineering: its role and function in human society. Elmsford, N.Y., Pergamon, 1967.

DeCamp, Lyon Sprague. The Ancient engineers. Cambridge, Mass., MIT Press, 1970, c1963.

Derry, Thomas Kingston. A Short history of technology, from the earliest times to A.D. 1900. New York, Oxford Univ. Pr., 1970.

Gablehouse, Charles. Helicopter and autogiros; a history of rotating-wing and V/STOL aviation. Rev. ed. Philadelphia, Lippincott, 1969.

Georgano, G. N. A History of sports cars. New York, Dutton, 1970.

Held, Robert. The Age of firearms; a pictorial history... rev. ed. Northfield, Ill., Gun Digest, 1970.

Hewlett, Richard G. & Anderson, Oscar E. Jr. New world, 1939 to 1946: a history of the United States Atomic Energy Commission. Univ. Park, Pa., Pa. State U. Pr., v. 1, 1962.

__________ & Duncan, Francis. Atomic shield, 1947 to 1952: a history of the United States Atomic Energy Commission. Univ. Park, Pa., Pa. State U. Pr., v. 2, 1969.

Hopkins, Henry James. A Span of bridges. New York, Praeger, 1970.

Miller, Ronald E. The Technical development of modern aviation. New York, Praeger, 1970, c1968.

Musson, Albert Edward. Science and technology in the industrial revolution. Toronto, Univ. of Toronto Press, 1969.

Nicholson, Timothy Robin. Passenger cars, 1863-1904. New York, Macmillan, 1970.

Pursell, Carroll W. Early stationary steam engines in America; a study in the migration of a technology. Washington, D.C., Smithsonian Inst. Pr., 1969.

Sloane, Howard N. & Lucille L. A Pictorial history of American mining. New York, Crown, 1970.

Smith, Cyril S., ed. Sources for the history of the science of steel, 1532-1786. Cambridge, Mass., MIT Press, 1968. (Society for the History of Technology Series, No. 4.)

Spence, Clark C. Mining engineers & the American West; the lace-boot brigade, 1849-1933. New Haven, Conn., Yale Univ. Press, 1970.

Straub, Hans. A History of civil engineering: an outline from ancient to modern times. London, Hill, 1952. Reprint: Cambridge, Mass., MIT Press, 1964. (Translated from 1949 German edition.)

Sweeney, James B. A Pictorial history of oceanographic submersibles. New York, Crown, 1970.

U.S. National Aeronautics and Space Administration. Astronautics and aeronautics; chronology on science, technology, and policy. Washington, D.C., 1963--. (Monthly with annual volumes.)

U.S. National Bureau of Standards. Measures for progress. Washington, D.C., U.S.G.P.O., 1966.

Von Braun, Wernher & Ordway, Frederick III. History of rocketry & space travel. Rev. ed. New York, T. Y.

Crowell, 1969.

Wherry, Joseph H. Automobiles of the world: the story of the development of the automobile with rare illustrations from a score of nations. Philadelphia, Chilton, 1968.

Wood, Peter & White, Edmund. When Zeppelins flew. New York, Time-Life, 1969.

Biographies

Well known individuals usually can be found listed in general encyclopedias, such as Encyclopedia Americana, Encyclopaedia Britannica, Collier's, or in the library's card catalog. A quick way to find a few facts about an important scientist is to look at the catalog cards for his books in the author/title section. Those catalog cards that have his name as author usually will furnish the following biographical information:

1. Dates of birth and death.
   These dates follow his name. If the author is dead, consult the various general encyclopedias and biographical dictionaries. However, if no death date is listed, assume he is still living and look for more information in such books as the appropriate Who's Who.

2. Nationality.
   Check the city where his books were published. The place of publication is given on the catalog card after the title of the book and often indicates the nationality of the author. To look up living Americans, use Who's Who in America. If the books were published in England, use the British biographical dictionary Who's Who.

3. Profession.
   Is he best known as a chemist, physicist, mathematician? There are special books you should try first for scientists. Often you can tell about the author's interests or profession from the titles and introductions to his books. Are they about ecology, oceanography, biochemistry? There are special biographical dictionaries for physicists, botanists, mathematicians and for other areas of science.

For additional biographical material, consult the library's catalog under such subject headings as:

Anatomists
Anthropologists

Astronomers
Biography - Dictionaries
Biologists
Botanists
Chemists
Engineers, American - Directories
Entomologists
Geologists
Mathematicians
Naturalists
Ornithologists
Paleontologists
Physicists
Science as a profession
Science teachers
Scientists - Biography (Q141-3)
Scientists, Catholic
Scientists, French, (German, etc.)
Scientists - Directories (Q145)
Speleologists
Women as scientists
Zoologists

For longer biographies of a person, look in the library's catalog under his name. Books which were written about him are listed there. Book-length biographies of scientists frequently contain a complete list of their publications. If a full length biography is not available, general and subject dictionaries may be helpful. Biographies of scientists are often included with information on the subject in other general reference books. Having made note of whatever information the card catalog or general reference books supplied, you are ready to look up the person in the general and special biographical dictionaries and other biographical reference books.

Biographical dictionaries are among the more widely used reference books to locate information about an individual. The difficulty lies in finding the name of a living person whose fame is current. Collected biographies have been published in nearly all subject areas. The listings may be brief or detailed; restricted to living or dead, or by country of birth. Transliterations and variations of spelling contribute to the difficulty in locating a name. Here are some of the most useful of the general biographical dictionaries:

For living Americans.

1. Biography Index. Bronx, N. Y., H. W. Wilson Co., v. 1-- 1946--

2. Current Biography Yearbook. Bronx, N. Y., H. W. Wilson Co., 1946-1968.

3. New Century Cyclopedia of Names. Ed. by Barnhart, C. L. & Halsey, W. D. New York, Appleton, 1969. 3 v.

4. Who's Who in America. Chicago, Marquis, v. 1-- 1899--

5. Who's Who of American Women. Chicago, Marquis, v. 1-- 1899--

For Americans no longer living.

1. Biography Index.

2. Dictionary of American Biography (DAB). Ed. by American Council of Learned Societies. New York, Scribners, 1928-1958. 11 v.

3. New Century Cyclopedia of Names.

4. Who Was Who in America. Chicago, Marquis, 1968. 4 v.

For people of other countries.

1. Dictionary of National Biography. New York, Oxford Univ. Pr., 1885-- 69 v.

2. International Who's Who, 1969-70. 33rd ed. New York, International Pubns. Service, 1969.

3. Webster's Biographical Dictionary. Springfield, Mass., Merriam Co., 1966.

Collected biographies, both historical and recent, supplement the general biographical dictionaries with information about persons connected with science or a particular scientific subject.

SCIENCE, GENERAL

American Council of Learned Societies Committee on Far Eastern Studies. Dictionary of scientific biography. New York, Scribner, 1970. 2 v.

American men of science, a biographical directory. 11th ed. New York, R. R. Bowker, 1965-67. 6 v. Supplements. 1966--.

Asimov, Isaac. Asimov's biographical encyclopedia of science & technology. New York, Doubleday, 1964.

A biographical dictionary of scientists. New York, Wiley, 1969.

Furth, Charlotte. Ting Wen-chiang; science and China's new culture. Cambridge, Mass., Harvard Univ. Press, 1970.

Haber, Louis. Black pioneers of science and invention. New York, Harcourt Brace, 1970.

McGraw-Hill Encyclopedia of Science & Technology Editors. Modern men of science. New York, McGraw-Hill, 1966-68. 2 v.

National Academy of Science. Biographical memoirs. Irvington-on-Hudson, N.Y., Columbia U. Pr., v. 1-- 1877-79--.

Royal Society of London. Biographical memoirs of fellows of the Royal Society. London, v. 1-- 1955--.

Turkevich, John & Ludmilla B., comp. Prominent scientists of continental Europe. New York, American Elsevier, 1968.

U.S. Department of State. Biographic register. 3 pts. Washington, D.C., U.S.G.P.O., 1869--. (title change 1942.)

Who's who in science in Europe: reference guide to West European scientists. New York, International Pubns. Service, 1968. 3 v.

World who's who in science; a biographical dictionary of

notable scientists from antiquity to the present... Chicago, Marquis Who's Who Inc., c1968.

MATHEMATICS

American Mathematical Society. Memoirs. Providence, Rhode Island, American Mathematical Society, v. 1, 1950--

Computer Consultant Editors. Who is related to whom in the computer industry. 3rd ed. Elmsford, N.Y., Pergamon, 1969.

ASTRONOMY

Geymonat, Ludovico. Galileo Galilei: a biography and inquiry into his philosophy of science. Tr. by Stillman Drake. New York, McGraw-Hill, 1965.

Kaplon, Morton F., ed. Homage to Galileo. Cambridge, Mass., M.I.T. Press, 1965.

PHYSICS

Andrade, E. N. da Costa. Sir Isaac Newton: his life and work. New York, Doubleday, 1965.

__________. Rutherford and the nature of the atom. New York Doubleday, 1964.

Cuny, Hilaire. Albert Einstein: the man and his theories. New York, Fawcett World, 1969.

DeLeeuw, Adele Louise. Marie Curie, woman of genius. Champaign, Ill., Garrard, 1970.

Ivimey, Alan. Marie Curie, pioneer of the atomic age. New York, Praeger, 1969.

McKnown, Robin. Marie Curie. New York, Putnam, 1971.

Nitske, Wilhelm Conrad. The Life of Wilhelm Conrad Rontgen, discoverer of the X ray. Tucson, Univ. of Arizona Press, 1971.

Who's who in atoms. 5th ed. New York, International Pubns. Service, 1969.

Williams, Leslie Pearce. Michael Faraday: a biography. New York, Basic Books, 1965.

CHEMISTRY

Cummings, Richard. The Alchemists: fathers of practical chemistry. New York, McKay, 1966.

Kuslan, Louis I. & Stone, A. Harris. Robert Boyle, the great experimenter. New York, Prentice-Hall, 1970.

Silverstein, Alvin & Virginia B. Harold Urey; the man who explored from earth to moon. New York, John Day, 1970, c1971.

BIOLOGICAL SCIENCES

Barnhart, John H., ed. Biographical notes upon botanists. Boston, Mass., G. K. Hall, 1965. 3 v.

Cuny, Hilaire. Louis Pasteur: the man and his theories. New York, Eriksson, 1966.

DeBeer, Gavin. Charles Darwin; evolution by natural selection. New York, Doubleday, 1964.

Huxley, Julian and Kettlewell, H. Bernard D. Charles Darwin and his world. New York, Viking, 1965.

MEDICAL SCIENCES

American men of medicine. 3rd ed. New York, Institute for Research in Biography, 1st ed., 1945--

Malgaigne, Joseph F. Surgery and Ambroise Paré. Tr. and ed. by Wallace B. Hamby. Norman, Okla., U. of O. Pr., 1965.

Stevenson, Lloyd G. Nobel prize winners in medicine and physiology: 1901-1964. Ed. by Theodore L. Sourkes. New York, Abelard, 1967.

Thacher, James. American medical biography. 2nd ed. New York, Plenum, 1967. 2 v.

__________ & Williams, Stephen W. American medical biography of 1845 & 1928. New York, Milford House, 1967. 2 v.

EINGINEERING SCIENCES

Adams, Jean & Kimball, Margaret. Heroines of the sky. Freeport, N. Y., Books for Libs., 1970, c1942.

Dwiggins, Don. Famous flyers and the ships they flew. New York, Grosset & Dunlap, 1969.

Evlanoff, Michael & Fluor, Marjorie. Alfred Nobel, the loneliest millionaire. Los Angeles, Ritchie, 1969.

Halacy, Daniel Stephen. Charles Babbage, father of the computer. New York, Crowell-Collier Pr., 1970.

Harris, Sherwood. The first to fly; aviation's pioneer days. New York, Simon & Schuster, 1970.

Lehman, Milton. This high man, the life of Robert H. Goddard. New York, Pyramid, 1970, c1963.

Lessing, Lawrence. Man of high fidelity: Edwin Howard Armstrong. New York, Bantam, 1969.

MacDonald, David Keith Chalmers. Faraday, Maxwell, and Kelvin. New York, Doubleday, 1964.

Thomson, George. J. J. Thomson: discoverer of the electron. New York, Doubleday, 1966. (Science Studies Series)

Webb, Robert N. James Watt, inventor of a steam engine. New York, Watts, 1970.

Who's who in engineering; a biographical dictionary of the engineering profession. New York, Lewis Historical Pub., 1st ed., 1922--, 9th ed., 1964.

Who's who of British engineers, 1968. Athens, Ohio, Ohio U. Pr., 1969.

For magazine articles about a person, consult the Applied Science and Technology Index. Biographical articles and obituaries appearing in science periodicals may also be found in the various indexing and abstracting services, either under the individual's name or under the heading "Biographies" or "Obituaries." They usually state whether the article is biographical, an obituary or a memorial lecture. Any of these may contain a comprehensive bibliography and should be checked as a source of science information.

For newspaper articles consult the New York Times Index under "Obituaries." The exact death date will establish the approximate time when stories might appear in other newspapers or periodicals.

The most up-to-date biographical notes of all are to be found in the news or personnel changes columns of science periodicals. Scientists working in academic institutions may be identified in brochures and pamphlets, together with details of any research programs. Another way of locating individuals working on particular problems is through their publishers. A mailing address of the author of a scientific article is usually printed with the article or editors may forward mail to their authors and contributors.

Directories

Information about authors who have written very little, or whose published work is very recent may be difficult to find. Those actively engaged in science usually belong to one or more of the national societies. The easiest way to identify them is through one of the society membership directories. Since questions concerning the names and addresses of scientists and professional societies may occur in a library search, knowing where to look is essential. Individual membership lists published by the societies or associations usually include the person's name, address, affiliation, publications and a brief personal history.

Information about the history, organization, officers, publications and addresses of the various learned societies can also be found in directories. Since thousands of directories are published, and no one library collects them all, often the searcher's main problem is locating the directory containing the appropriate listing. Directories can vary from records of current developments in the subject to statistical compilations. Some of the more useful directories of persons, institutions, firms and societies in science and technology follow.

Directories of libraries of specific countries, or of special libraries or collections in particular fields are available. Many of these directories, both of the U. S. and of other countries, will help students to locate libraries which house special collections or which are strong in particular subjects. In some directories library holdings are described in detail, in others mere listings of outstanding collections are given.

SCIENCE, GENERAL

American Chemical Society. Directory of graduate research. Washington, D. C., American Chemical Society, v. 1, 1953-- (Biennial. Title varies.)

American Translators Association. ATA professional services directory. New York, 1965--

Directory of British scientists. London, Benn, 1963-- (Annual.)

Directory of computerized information in science & technology. New York, Science Associates International, 1968. (Looseleaf with supplements.)

Directory of Department of Defense information analysis centers. Washington, D. C., U. S. G. P. O., 1966.

Directory of federally supported information analysis centers. Springfield, Va., Clearinghouse, 1968.

Directory of national trade and professional associations of the United States. Washington, D. C., Columbia Books, 1966-1970. 5 v.

Directory of selected research institutes in Eastern Europe. Irvington-on-Hudson, N. Y., Columbia Univ. Press, 1967.

European Research Index. 2nd ed. Ed. by Williams, C. H. New York, International Pubns. Service, 1969. 2 v.

Gale Research Co. Directory of special libraries and information centers. Ed. by Kruzas, A. T. 2nd ed. Detroit, Gale Research Co., 1969. 2 v.

__________. Encyclopedia of associations. 5th ed. Ed. by Ruffner, F. G. Detroit, Gale, 1968. 3 v.

__________. Research centers directory. 3rd ed. Ed. by Palmer, Archie. Detroit, Gale, 1968.

Guide to European sources of technical information. 3rd ed. New York, International Pubns. Service, 1969.

Industrial research laboratories of the United States. 13th ed. New York, R. R. Bowker, 1970. (Earlier editions in the NAS-NRC Publication series.)

MacKeigan, Helaine, ed. American Library Directory, 1970-71. 27th ed. New York, R. R. Bowker, 1971.

MacRae's blue book. Philadelphia, Macrae Smith Co., v. 1, 1881-- (Title varies.)

Museums directory of the United States and Canada. 2nd ed.

Washington, D. C., American Assn. of Museums and Smithsonian Institution, 1965.

National Academy of Sciences. Scientific & technical societies of the U. S. 8th ed. Washington, D. C., Nat'l. Acad. of Sci., 1968. (NAS-NRC Pub. 1499.)

Neverman, F. John, ed. International directory of back issue vendors: periodicals, newspapers, newspaper indexes, documents. 2nd enl. ed. New York, Spec. Libraries Assn., 1968.

Scientific meetings. New York, Spec. Libraries Assn., v. 1, 1957--

Technical Meetings Information Service. World meetings, outside U. S. A. and Canada. Newton Centre, Mass., 1968--

__________. World meetings, U. S. and Canada. Newton Centre, Mass., v. 1, 1963-- (Title change with v. 5, April 1966.)

Thomas' register of American manufacturers. New York, Thomas, 1906-- (Annual.) 5 v.

UNESCO. World directory of national science policy-making bodies. v. 1: Europe and North America. New York, 1966-68. 3 v.

__________. World guide to technology information and documentation services. New York, Unipub, 1970.

U. S. Government organization manual. Washington, D. C., U. S. G. P. O., 1935-- (Title varies.)

U. S. Library of Congress. Directories in science and technology: a provisional checklist. Washington, D. C., U. S. G. P. O., 1964.

__________. World list of future international meetings. Washington, D. C., 1959--

U. S. National Referral Center. A Directory of information resources in the United States: Federal Government. Washington, D. C., U. S. G. P. O., 1967.

__________. A Directory of information resources in the United States: physical sciences, biological sciences, engineering. Washington, D. C., U. S. G. P. O., 1965.

Watkins, Ralph J. Directory of selected scientific institutions in Mainland China. Stanford, Cal., Hoover Institution, 1971.

Wynkoop, Sally & Parish, David W., comps. Directories of government agencies. Rochester, N. Y., Libraries Unlimited, 1969.

Yugoslav scientific research directory. Springfield, Va., Clearinghouse, 1964.

MATHEMATICS

American Mathematical Society, Mathematical Association of America, and Society for Industrial & Applied Mathematics. Combined membership list. Buffalo, N. Y., SUNY. Annual.

International directory of computer & information system services, 1969. New York, International Pubns. Service, 1969.

World directory of mathematicians. Utrecht, International Mathematical Union, 1958--

CHEMISTRY

Chem sources. Flemington, N. J., Directories Pub. Co., 1958--

Chemical guide to Europe. Pearl River, N. Y., Noyes Development Co., 1963--

Chemical guide to the United States. Pearl River, N. Y., Noyes Development Co., 1962--

Chemical materials catalog and directory of producers. New York, Van Nostrand-Reinhold, 1949-50--

International chemistry directory. New York, W. A. Benjamin, 1969--

An Overview of worldwide chemical information facilities and resources. Springfield, Va., Clearinghouse, 1968. (PB176-160.)

Zimmerman, O. T. & Lavine, I. Handbook of material trade names. 2nd ed. Dover, N.H., Industrial Research Service, 1953. Supplements, 1956--

EARTH SCIENCES

Directory of oceanographers in the United States. Washington, D.C., National Academy of Sciences, 1969.

Geophysical directory. Houston, Texas, v. 1, 1946--

National Research Council. Oceanography information sources. Washington, D.C., NAS-NRC, 1966. (NAS-NRC Pub. 1917.)

UNESCO. Directory of meteorite collections and meteorite research. New York, 1968.

BIOLOGICAL SCIENCES

American Institute of Biological Sciences, comp. Directory of bioscience departments in the United States and Canada. New York, Van Nostrand-Reinhold, 1967.

Mallis, Arnold. American entomologists. New Brunswick, N.J., Rutgers Univ. Pr., 1971.

MEDICAL SCIENCES

American Dental Association. American dental directory. Chicago, 1st ed., 1947--

American Hospital Association. Hospitals: guide issue. Chicago, 1945--

American Medical Association. American medical directory. Chicago, 1st ed., 1906-- 25th ed., 1969, 3 v.

__________. Directory of approved internships and residencies. Chicago, 1947--

Directory of health science libraries in the United States. Chicago, American Medical Association, 1969.

Directory of medical specialists. v. 14. Chicago, Marquis Who's Who, Inc., 1970.

Directory of medical specialists holding certification by American Specialty Boards. Chicago, Advisory Board of Medical Specialties, 1st ed., 1939--

Drugs in current use. New York, Springer Pub., 1970.

United Nations. Drug Supervisory Body. Estimated world requirements of narcotic drugs and estimates of world production of opium. New York, United Nations, 1970.

U.S. Public Health Service. Bureau of State Services. Directory of state and territorial health authorities. Washington, D.C., U.S.G.P.O., 1913-- (Annual.)

AGRICULTURAL SCIENCES

American Veterinary Medical Association. Directory. Chicago, 1st ed., 1924--

Blase, Melvin G., ed. Institutions in agricultural development. Ames, Iowa, Iowa State Univ. Press, 1971.

Directory of agricultural and home economic leaders of the U.S. and Canada. Cambridge, Mass., W. G. Grant, 1919-- (Annual.)

IFI-Plenum Data Corporation. Directory, 1965-68. New York, 1968. (Looseleaf; annual revision service.)

U.S. Department of Agriculture. Directory of organizations and field activities of the Department of Agriculture. Washington, D.C., U.S.G.P.O., v. 1, 1925--

__________. Workers in subjects pertaining to agriculture in the land-grant colleges and experiment stations... Washington, D.C., U.S.G.P.O., 1924-25--

ENGINEERING SCIENCES

Allen, Roy. Great airports of the world. 2nd ed. New Rochelle, N.Y., Sportshelf, 1968.

American Society for Testing and Materials. Directory of testing laboratories, commercial-institutional. Philadelphia, ASTM, 1970.

Engineers Joint Council. Directory of engineering societies and related organizations. New York, 1968. (Formerly Engineering Societies Directory, 1948--.)

Jones, Vane A. North American radio-TV station guide. 6th ed. Indianapolis, Ind., Sams, 1970.

National Research Council. Committee on Fire Research. Directory of fire research in the United States. 5th ed. Washington, D.C., National Academy of Sciences, National Research Council, 1967-1969.

National Society of Professional Engineers. Directory of engineers in private practice. Washington, D.C., 1965-- (Annual.)

Rolfe, Douglas. Airplanes of the world. 3rd rev. & enl. New York, Simon & Schuster, 1969.

U.S. Public Health Service. Directory of homemaker services: homemaker agencies in the United States. Washington, D.C., U.S.G.P.O., 1958--

Vanderveen, Bart Harmannus. The Observer's fighting vehicles directory, World War II. New York, Warne, 1969.

World aviation directory. Washington, D.C., American Aviation Pub., v. 1, 1940--. (Semiannual.)

World space directory. Washington, D.C., American Aviation Pub., v. 1, 1963--. (Semiannual.)

Handbooks and Manuals

Almost as useful as the subject encyclopedia in a library search is the handbook or manual. Usually written about one subject, handbooks supply charts, formulas, tables, statistical data, and historical background. They are revised frequently, sometimes yearly, to include new developments and practices. Handbooks are comprehensive in scope, condensed in treatment and arranged on a classified or systematic plan to help find information quickly using the table of contents.

Single definite facts can usually be located through a detailed alphabetical index. In science and technology there are handbooks which are especially useful because they contain bibliographies or footnote references to additional information.

SCIENCE, GENERAL

Boy Scouts of America. Fieldbook for boy scouts, explorers, scouters, educators & outdoorsmen. New York, Boy Scouts of America, 1967.

CRC composite index for CRC handbooks. Cleveland, Chemical Rubber Co., 1971.

Fisher, Ronald A. and Yates, Frank. Statistical tables for biological, agricultural and medical research. 6th ed. New York, Hafner, 1964.

Fuller, Richard Buckminster. Operating manual for spaceship earth. New York, Pocket Book, 1970, c1969.

Gabb, M. H. & Latcham, W. E. Handbook of laboratory solutions. New York, Chem. Pub., 1968.

Gray, Peter. Handbook of basic microtechnique. 3rd ed. New York, McGraw-Hill, 1964.

Kaye, George W. & Laby, T. H., comp. Tables of physical & chemical constants. 13th ed. New York, Wiley, 1966.

National Research Council -- Office of Critical Tables. Continuing numerical data projects: a survey & analysis. 2nd ed. Washington, D.C., National Academy of Sciences, 1966.

Needham, George Herbert. The Microscope; a practical guide. Springfield, Ill., C. C. Thomas, 1968.

Stehl, Georg Jakob. The Microscope and how to use it. New York, Dover, 1970, c1960.

Weast, R. C. and Selby, S. M. CRC handbook of chemistry and physics. 52nd ed. Cleveland, Chemical Rubber Co., 1971.

MATHEMATICS

Abramowitz, Milton and Stegun, Irene A., eds. Handbook of mathematical functions with formulas, graphs and mathematical tables. New York, Dover, 1964.

Alger, Philip L. Mathematics for science & engineering. 2nd ed. New York, McGraw-Hill, 1969.

Arfken, George Brown. Mathematical methods for physicists. 2nd ed. New York, Academic Press, 1970.

Benjamin, Jack R. & Cornell, C. Allin. Probability, statistics, and decisions for civil engineers. New York, McGraw-Hill, 1970.

Beyer, William H. CRC handbook of tables for probability and statistics. 2nd ed. Cleveland, Chemical Rubber Co., 1968.

Burington, Richard Stevens. Handbook of mathematical tables and formulas. 4th ed. New York, McGraw-Hill, 1965.

__________ & May, Donald C., Jr. Handbook of probability and statistics with tables. 2nd ed. New York, McGraw-Hill, 1970.

CRC handbook of tables for mathematics. 4th ed. Cleveland, Chemical Rubber Co., 1970.

Cameron, Archibald James. A Guide to graphs. Elmsford, N. Y., Pergamon, 1970.

Chatfield, Christopher. Statistics for technology. Baltimore, Penguin, 1970.

Clarke, A. Bruce & Disney, Ralph L. Probability and random processes for engineers and scientists. New York, Wiley, 1970.

Cooke, Nelson M. and Adams, Herbert F. R. Basic mathematics for electronics. 3rd ed. New York, McGraw-Hill, 1970.

Drooyan, Irving. Manual for the slide rule. 2nd ed. Belmont, Cal., Wadsworth Publishing, 1970.

Fisher, Sir Ronald Aylmer. Statistical methods for research workers. 14th ed. rev. & enl. Darien, Conn., Hafner, 1970.

Fitzgerald, William M. Laboratory manual for elementary mathematics. Boston, Prindle, Weber, 1969.

Fox, Leslie and Mayers, D. F. Computing methods for scientists and engineers. New York, Oxford Univ. Press, 1968.

Grau, A. A., et al. Handbook for automatic computation, v. 1, pt. B: translation of ALGOL 60. New York, Springer-Verlag, 1967. (Grundlehren der Mathematischen Wissenschaften, v. 137.)

Gray, Henry L. & Odell, Patrick L. Probability for practicing engineers. New York, Barnes & Noble, 1970.

Grazda, Edward E., et al. Handbook of applied mathematics. 4th ed. New York, Van Nostrand-Reinhold, 1966.

Harrington, Rodger F. Field computation by moment methods. New York, Macmillan, 1968.

Heading, John. Mathematical methods in science and engineering. 2nd ed. New York, American Elsevier, 1970.

Horton, Holbrook L. Machinery's mathematical tables. 3rd ed. New York, Industrial Press, 1969.

Jeffrey, Alan. Mathematics for engineers and scientists. New York, Barnes & Noble, 1970.

Kelly, L. G. Handbook of numerical methods & applications. Reading, Mass., Addison-Wesley, 1967.

Klerer, M. & Korn, G. A. Digital computer user's handbook. New York, McGraw-Hill, 1967.

Korn, Granino Arthur and Theresa M. Mathematical handbook for scientists and engineers; definitions, theorems and formulas for reference and review. 2nd ed., New York, McGraw-Hill, 1967.

Krylov, V. I. & Skoblya, N. S. Handbook of numerical inversion of Laplace transforms. (Tr. by Louvish, D.) Hartford, Conn., Davey, 1969.

Liebeck, Hans. Algebra for scientists and engineers. New York, Wiley, 1969.

Lyusternik, L. A., et al. Handbook for computing elementary functions. Elmsford, N.Y., Pergamon, 1965.

Mann, Lawrence. Applied engineering statistics for practicing engineers. New York, Barnes & Noble, 1970.

Margenau, Henry & Murphy, George M. Mathematics of physics & chemistry. 2nd ed. New York, Van Nostrand-Reinhold, 1964. 2 v.

Marriott, Francis Henry Charles. Basic mathematics for the biological and social sciences. Elmsford, N.Y., Pergamon, 1970.

Martin, Hedley G. Mathematics for engineering, technology, and computing science. Elmsford, N.Y., Pergamon, 1970.

Masterton, William L. Mathematical preparation for general chemistry. Philadelphia, Saunders, 1970.

Mathews, Jon & Walker, R. L. Mathematical methods of physics. 2nd ed. New York, W. A. Benjamin, 1970.

Merritt, Frederick S. Applied mathematics in engineering practice. New York, McGraw-Hill, 1970.

__________. Modern mathematical methods in engineering. New York, McGraw-Hill, 1970.

Nixon, Floyd E. Handbook of Laplace transformation. 2nd ed. New York, Prentice-Hall, 1965.

Perrin, Charles L. Mathematics for chemists. New York, Wiley (Interscience), 1970.

Remington, Richard Delleraine & Schork, M. Anthony. Statistics with applications to the biological and health sciences. New York, Prentice-Hall, 1970.

Rutishauser, H. Handbook for automatic computation, v. 1, pt. A: translation of ALGOL 60. New York, Springer-Verlag, 1967. (Grundlehren der Mathematischen Wissenschaften, v. 135.)

Schefler, William C. Statistics for the biological sciences. Reading, Mass., Addison, 1969.

Tuma, Jan J. Engineering mathematics handbook: definitions, theorems, formulas, tables. New York, McGraw-Hill, 1970.

Volk, William. Applied statistics for engineers. 2nd ed. New York, McGraw-Hill, 1969.

Walsh, John E. Handbook of non parametric statistics. New York, Van Nostrand-Reinhold, 1962-1968. 3 v.

Westlake, Joan R. Handbook of numerical matrix inversions and solution of linear equations. New York, Wiley, 1968.

ASTRONOMY

Alter, Dinsmore. Pictorial guide to the moon. Rev. ed. New York, T. Y. Crowell, 1967.

__________, et al. Pictorial astronomy. 3rd ed. New York, T. Y. Crowell, 1969.

Celestial handbook. Washington, D. C., Celestial Press, 1966--

Cherrington, E. H., Jr. Exploring the moon through binoculars. New York, McGraw-Hill, 1969.

Howard, Neale E. The Telescope handbook and star atlas. New York, T. Y. Crowell, 1967.

Jackson, Joseph H. Pictorial guide to the planets. New York, T. Y. Crowell, 1965.

King, Henry C. Pictorial guide to the stars. New York, T. Y. Crowell, 1967.

Menzel, Donald H. Field guide to the stars and planets. Boston, Houghton-Mifflin, 1964.

Moore, Patrick, ed. Handbook of practical amateur astronomy. New York, Norton, 1964.

Muirden, James. Amateur astronomer's handbook. New York, T. Y. Crowell, 1968.

Neely, Henry Milton. A Primer for star-gazers. New York, Harper, 1970.

Page, T. & L. Telescopes; how to make them and use them. New York, Macmillan, 1966.

Roth, Gunter Dietmar. Handbook for planet observers. New York, Van Nostrand-Reinhold, 1970.

Shaw, R. William and Boothroyd, Samuel L. Manual of astronomy. 5th ed. Dubuque, Iowa, Wm. C. Brown, 1967.

U.S. Nautical Almanac Office. American ephemeris and nautical almanac. Washington, D.C., U.S.G.P.O., 1855--. (Annual.)

PHYSICS

American Institute of Physics. American Institute of Physics handbook. Ed. by Dwight E. Gray. 3rd ed. New York, McGraw-Hill, 1970.

Condon, Edward W. & Odishaw, H. Handbook of physics. 2nd ed. New York, McGraw-Hill, 1967.

Ebert, Hermann, ed. Physics pocketbook. New York,

Wiley (Interscience), 1967.

Heard, H. G. Laser parameter measurement handbook. New York, Wiley, 1968.

Kaelble, E. F. Handbook of X-rays: for diffraction, emission, absorption and microscopy. New York, McGraw-Hill, 1967.

Koshkin, N. & Shirkevich, M. Handbook of elementary physics. (Tr. by F. Leib), New York, Gordon & Breach, 1965.

Kunz, W. & Schintlmeister, J. Nuclear tables. Elmsford, N.Y., Pergamon, 1958-1968. 3 v.

Nuclear data. New York, Academic Press, Section A, 1965-- Section B, 1966-- (Supersedes National Research Council's Nuclear Data Sheets.)

Pressley, Robert J. CRC handbook of lasers with selected data on optical technology. Cleveland, Chemical Rubber Co., 1971.

Szymanski, Herman A. & Erickson, Ronald E. Infrared band handbook. 2nd ed. rev. & enl. New York, Plenum (IFI), 1970. 2 v.

Williams, Richard A. Handbook of the atomic elements. New York, Philosophical Lib., 1970.

CHEMISTRY

Brauer, G., ed. Handbook of preparative inorganic chemistry. 2nd ed. New York, Academic Press, 1963-1965. 2 v.

Chayen, Joseph. Guide to practical histochemistry. Philadelphia, Lippincott, 1969.

Dawson, John W. Laboratory manual for chemistry: a brief introduction. Philadelphia, Saunders, 1970.

Handbook for chemistry assistants. Easton, Pa., Chemistry Education, 1965.

Hirayama, Kenzo. Handbook of ultraviolet and visible absorption spectra of organic compounds. New York, Plenum (IFI), 1967.

Keller, Roy A. Basic tables in chemistry. New York, McGraw-Hill, 1967.

Lajtha, Abel, ed. Handbook of neurochemistry. New York, Plenum, v. 3 & 4, 1970. (proposed 7 v.)

Lange, Norbert A. Handbook of chemistry. 10th ed. New York, McGraw-Hill, 1967.

Latimer, Wendall M. & Hildebrand, Joel H. Reference book of inorganic chemistry. 3rd ed. New York, Macmillan, 1964.

National Research Council. Specifications and criteria for biochemical compounds. 2nd ed. Washington, D.C., NAS-NRC, 1967. (NAS-NRC Pub. No. 1344.)

Polymer handbook. New York, Wiley, 1966.

Rappoport, Zvi, ed. CRC handbook of tables for organic compound identification. 3rd ed. Cleveland, Chemical Rubber Co., 1967.

Ruben, Samuel. Handbook of the elements. 2nd ed. Indianapolis, Ind., Sams, 1967.

Samsonov, Grigory V. Handbook of the physicochemical properties of the elements. New York, Plenum (IFI), 1968.

Say, N. I., ed. Dangerous properties of industrial materials. 3rd ed. New York, Van Nostrand-Reinhold, 1968.

Steeve, Norman V., ed. CRC handbook of laboratory safety. 2nd ed. Cleveland, Chemical Rubber Co., 1970.

Sunshine, Irving, ed. CRC handbook of analytical toxicology. Cleveland, Chemical Rubber Co., 1971.

__________. CRC manual of analytical toxicology. Cleveland, Chemical Rubber Co., 1971.

Sykes, Peter. A Guidebook to mechanism in organic chemistry. 3rd ed. New York, Wiley, 1970.

Szymanski, H. A. & Yelin, R. E. NMR band handbook. New York, Plenum, 1968.

Weiss, Howard D. Guide to organic reactions. Minneapolis, Burgess, 1969.

White, Robert G. Handbook of ultraviolet methods. New York, Plenum, 1965.

EARTH SCIENCES

Chow, Ven-Te, ed. Handbook of applied hydrology; a compendium of water resources technology. New York, McGraw-Hill, 1964.

Dana, James Dwight. Dana's manual of mineralogy. 18th ed. New York, Wiley, 1971.

Denny, Charles S., et al. A Descriptive catalog of selected aerial photographs of geologic features of the United States. Washington, D.C., U.S.G.P.O., 1968. (U.S. G.S. Professional Paper 590.)

Filyand, M. A. and Semenova, E. I. Handbook of rare elements. Ed. by Alferiett, M. E. v. 1, trace elements; v. 2, refractory elements; v. 3, radioactive elements. Cambridge, Mass., Boston Tech., 1968.

Handbook of oceanographic tables, 1966. Washington, D.C., U.S.G.P.O., 1967.

Johnson, Paul W. Field guide to the gems and minerals of Mexico. Ed. by MacLachlan. Mentone, Cal., Gembooks, 1965.

Kummel, Bernhard & Raup, David, eds. Handbook of paleontological techniques. San Francisco, W. H. Freeman, 1965.

Lefond, Stanley J. Handbook of world salt resources. New York, Plenum, 1969. (Monographs in Geoscience Series.)

Liddicoat, Richard T., Jr. Handbook of gemstone identification. 8th ed., Evanston, Ill., Gemological, 1966.

Life Editors. Handbook of the nations. New York, Time-Life, 1966.

List, Robert J. Smithsonian meteorological tables. rev. ed. New York, Random (Smithsonian), 1966.

Lyon, Edward E. Earth science manual. 3rd ed. Dubuque, Iowa, W. C. Brown, 1971, c1964.

MacFall, R. P. Gem hunter's guide: how to find and identify gem minerals. 4th ed., New York, T. Y. Crowell, 1969.

Matthews, William H. A Guide to the national parks, their landscape and geology. v. 1, western parks; v. 2, eastern parks. New York, Doubleday (Natural History), 1968. 2 v.

Mirkin, Lev I. Handbook of X-ray analysis of polycrystalline materials. New York, Plenum (IFI), 1964.

Myers, John J., ed. Handbook of ocean and underwater engineering. New York, McGraw-Hill, 1969.

Nicholson, J. R. Meterological data catalogue. Honolulu, East-West Center Press, 1969.

Pohopien, K. M. Field manual: an introduction to the megascopic study & determination of minerals & rocks. Dubuque, Iowa, Wm. C. Brown, 1969.

Ransom, Jay E. A Range guide to mines and minerals: how and where to find valuable ores and minerals in the U.S. New York, Harper & Row, 1964.

Sanborn, William B. Handbook of crystal & mineral collecting. Mentone, Cal., Gembooks, 1966.

Tomikel, John. Handbook of minerals. California, Pa., Allegheny Press, 1968.

UNESCO. Manual on International Oceanographic Data Exchange. 2nd ed. New York, Unipub, 1967. (Intergovernmental Oceanographic Commission. Technical Series, v. 4.)

U.S. Coast and Geodetic Survey. United States earthquakes.

Washington, D.C., U.S.G.P.O., 1930--

U.S. Weather Bureau. Climates of the states. Washington, D.C., U.S.G.P.O., 1959-- (Climatology of the United States No. 60.)

_________. Climatological data for the United States by sections. Washington, D.C., U.S.G.P.O., v. 1-- 1914--. (monthly with annual summaries.)

_________. Climatological data; national summary. Washington, D.C., U.S.G.P.O., v. 1-- 1950--. (monthly with annual summary.)

_________. World weather records. Washington, D.C., U.S.G.P.O., v. 4-- 1959--. (v. 3 issued by Smithsonian Institution.)

Uytenbogaardt, W. Tables for microscopic identification of ore minerals. Princeton, N.J., Princeton Univ. Pr., 1951. Reprint: New York, Hafner, 1968.

Vanders, Iris & Kerrs, Paul F. Mineral recognition. New York, Wiley, 1967.

Von Bernewitz, W. M. Handbook for prospectors and operators of small mines. New York, McGraw-Hill, 1969.

Wedepohl, K. H., et al, eds. Handbook of geochemistry. New York, Springer-Verlag, 1968. 2 v.

BIOLOGICAL SCIENCES

Allen, B. L. Basic anatomy; a laboratory manual. San Francisco, W. H. Freeman, c1970.

Armstrong, George G. A Laboratory manual for Guyton's function of the human body. 2nd ed. Philadelphia, Saunders, 1969.

Austin, Elizabeth S. & Oliver L., Jr. Random House book of birds. New York, Random House, 1970.

Baker, Francis J. Handbook of bacteriological techniques. 2nd ed. New York, Appleton, 1967.

Barbour, Roger W. & Davis, Wayne H. Bats of America. Lexington, Ky., Univ. of Ky. Pr., 1970.

Benton, Allen H. & Werner, William E. Manual of field biology and ecology. Minneapolis, Burgess, 1965.

Bishop, Sherman C. Handbook of salamanders: the salamanders of the United States, of Canada, and of Lower California. Ithaca, N. Y., Comstock, 1943 & 1967.

Blair, W. F., et al. Vertebrates of the United States. 2nd ed. New York, McGraw-Hill, 1968.

Blake, Sidney F. Geographical guide to floras of the world. Washington, D. C., U. S. G. P. O., 1942-- (v. 1, 1942; v. 2, 1960.)

Bliss, Chester I. Statistics in biology, vol. 1. New York, McGraw-Hill, 1967.

Borror, Donald Joyce. A Field guide to the insects of America north of Mexico. Boston, Houghton-Mifflin, 1970.

Bridges, William. The New York aquarium book of the water world; a guide to representative fishes, aquatic invertebrates, reptiles, birds, and mammals. New York, N. Y. Zoological Society, 1970.

Burt, William H. and Grossenheider, R. P. Field guide to the mammals. 2nd ed. Boston, Houghton-Mifflin, 1964.

Calderone, Mary Steichen, ed. Manual of family planning and contraceptive practice. 2nd ed. Baltimore, Williams & Wilkins, 1970.

Campbell, R. C. Statistics for biologists. New York, Cambridge Univ. Pr., 1967.

Carlander, Kenneth Dixon. Handbook of freshwater fishery biology. vol. 1. 3rd ed. Ames, Iowa, Iowa St. U. Pr., 1969.

Cochran, Doris Mabel. The New field book of reptiles and amphibians. New York, Putnam, 1970.

Collins, Henry H. Jr. & Boyajian, Ned R. Familiar garden

birds of America. New York, Harper & Row, 1965. (Familiar Nature Series.)

Crouch, James E. Text-atlas of cat anatomy. Philadelphia, Lea & Febiger, 1969.

Dallimore, W. & Jackson, A. B. Handbook of coniferae & ginkgoaceae. 4th ed. New York, St. Martin, 1967.

Datta, Subhash Chandra. A Handbook of systematic botany. 2nd ed. New York, Asia Pub., 1970, c1965.

DeSachauensee, Rodolphe Meyer. A Guide to the birds of South America. Wynnewood, Pa., Livingston Pub. for Academy of Natural Sciences of Philadelphia, 1970.

Ditmars, Raymond L. Snakes of the world. New York, Macmillan, 1966.

Duddington, C. L. Beginner's guide to botany. London, British Book Centre, 1971.

Eddy, Samuel. How to know the freshwater fishes. 2nd ed. Dubuque, Iowa, Wm. C. Brown, 1970.

Ennion, Eric A. & Tinbergen, Nikolaas. Tracks. New York, Oxford Univ. Pr., 1968.

Everett, Thomas H. Living trees of the world. New York, Doubleday, 1969.

Faulkner, Willard R., ed. CRC handbook of clinical laboratory data. 2nd ed. Cleveland, Chemical Rubber Co., 1968.

__________. CRC manual of clinical laboratory procedures. 2nd ed. Cleveland, Chemical Rubber Co., 1970.

Fosberg, F. R. and Sachet, M. Manual for tropical herbaria. New York, S-H Service Agency, 1965.

Gersh, Harry. The Animals next door; a guide to zoos and aquariums of the Americas. New York, Fleet Academic, 1971.

Gibbs, Brian M. & Skinner, Frederick A., eds. Identification methods for microbiologists. New York, Academic

Press, 1966-1968. 2 pts.

Gilbert, Stephen G. Pictorial anatomy of the cat. Seattle, Univ. of Wash. Pr., 1968.

_________. Pictorial anatomy of the frog. Seattle, Univ. of Wash. Pr., 1965.

Gimble, Frank. Physiology; laboratory manual. Dubuque, Iowa, Wm. C. Brown, 1969.

Gosner, Kenneth L. Guide to identification of marine and estaurine invertebrates: Cape Hatteras to the Bay of Fundy. New York, Wiley (Interscience), 1971.

Gray, Peter. Use of the microscope; an introduction handbook for biologists. New York, McGraw-Hill, 1967.

Grimm, William C. Familiar trees of America. New York, Harper & Row, 1967.

_________. Recognizing flowering wild plants. Harrisburg, Pa., Stackpole, 1968.

_________. Home guide to trees, shrubs and wild flowers. Harrisburg, Pa., Stackpole, 1970.

Grossman, Mary L. & Hamlet, J. Birds of prey of the world. New York, Potter, 1970.

Guttmacher, Alan Frank. Understanding sex; a young person's guide. New York, New American Library, 1970.

Harrison, Bruce M. Manual of comparative anatomy, a general laboratory guide. St. Louis, Mosby, 1970.

Hazlewood, Walter G. Handbook of trees, shrubs & roses. Rev. ed. San Francisco, Tri-Ocean, 1969.

Headstrom, Birger Richard. A Complete field guide to nests in the United States, including those of birds, mammals, insects, fishes, reptiles and amphibians. New York, Washburn, 1970.

Higgins, Lionel George & Riley, Norman D. A Field guide to the butterflies of Britain and Europe. Boston, Houghton-Mifflin, 1970.

Howell, Trevor H. A Student's guide to geriatrics. 2nd ed. Springfield, Ill., C. C. Thomas, 1970.

Hughes, Helena E. & Dodds, T. C. Handbook of diagnostic cytology. Baltimore, Williams & Wilkins, 1968.

Hutchinson, John. Key to the families of flowering plants of the world. New York, Oxford Univ. Pr., 1967.

Ingles, L. G. Mammals of the Pacific states: California, Oregon, and Washington. Stanford, Cal., Stanford Univ. Pr., 1965.

Johnsgard, Paul A. Handbook of waterfowl behavior. Ithaca, N.Y., Comstock, 1965.

King, Lawrence J. Weeds of the world; biology and control. v. 1, New York, Wiley, 1966.

Kingsbury, John M. Deadly harvest; a guide to common poisonous plants. New York, Holt Rinehart & Winston, 1965.

__________. Poisonous plants of the United States & Canada. 3rd ed. New York, Prentice-Hall, 1964.

Lemon, Paul C. General botany manual; exercises on the life histories, structures, physiology, and ecology of the plant kingdom. 3rd ed. St. Louis, Mosby, 1970.

Lutz, Earnest L. Handbook of plastic embedding of animals, plants & various objects with improved C.M.E. D6 polyester resins. Ed. by Brown, Vinson. Healdsburg, Cal., Naturegraph, 1970.

McCluney, William Ross, ed. The Environmental destruction of South Florida: a handbook for citizens. Miami, Univ. of Miami Pr., 1971.

MacInnis, Austin J. Experiments and techniques in parasitology. San Francisco, Freeman, c1970.

McKenny, Margaret and Peterson, Roger Tory. Field guide to wildflowers of Northeastern and North-Central North America. Boston, Houghton-Mifflin, 1968. (Peterson Field Guide Series.)

Mathews, Ferdinand Schuyler. Field book of wild birds and their music. New York, Dover, 1966.

Merchant, Donald J., et al. Handbook of cell & organ culture. Minneapolis, Burgess, 1964.

Milne, Lorus & Margery. North American birds. New York, Prentice-Hall, 1969.

Morris, Desmond. The Mammals: a guide to the living species. New York, Harper & Row, 1966.

Morris, Percy A. Field guide to the shells of the Pacific coast and Hawaii. Rev. ed. Boston, Houghton-Mifflin, 1966.

Napier, J. R. & P. H. Handbook of living primates. New York, Academic Press, 1967.

Nelson, Ruth Elizabeth. Handbook of Rocky Mountain plants. Tucson, Ariz., D. S. King, 1969.

New York Times. New York Times book of trees & shrubs. Ed. by Faust, J. L., New York, Knopf, 1964.

Peter, Richard E. A Student's guide to laboratory experiments in general and comparative endocrinology. New York, Prentice-Hall, 1970.

Peterson, Roger Tory. Field guide to the birds. Boston, Houghton-Mifflin, 1968.

Polunin, Oleg. Flowers of Europe: a field guide. New York, Oxford Univ. Pr., 1969.

Prevot, Andre R. Manual for the classification and determination of the anaerobic bacteria. Tr. by Fredette, V. Philadelphia, Lea & Febiger, 1966.

Raffauf, Robert Francis. A Handbook of alkaloids and alkaloid-containing plants. New York, Wiley, 1970.

Rand, Austin L. & Gillard, E. Thomas. Handbook of New Guinea birds. New York, Doubleday (AMS Natural History), 1968.

Reisigl, Herbert, et al, eds. The world of flowers. New

York, Viking, 1966.

Rickett, Harold William. Wild flowers of the United States. New York, McGraw-Hill, 1966-1969. 3 v.

Ride, W. D. L. A Guide to the native mammals of Australia. New York, Oxford Univ. Pr., 1970.

Robbins, Chandler S., et al. Birds of North America; a guide to field identification. New York, Golden Press, 1968.

Rue, Leonard L. Pictorial guide to the birds of North America. New York, T. Y. Crowell, 1970.

__________. Pictorial guide to the mammals of North America. New York, T. Y. Crowell, 1967.

__________. Sportsman's guide to game animals. New York, Harper & Row, 1968.

Saunders, John Tennent. A Manual of practical vertebrate morphology. 4th ed. rev. Oxford, Clarendon Press, 1969.

Short, Douglas J., ed. The I. A. T. manual of laboratory animal practice & techniques. (Institute of Animal Technicians.) 2nd ed. London, Lockwood, 1969.

Silvan, James. Raising laboratory animals; a handbook for biological and behavioral research. New York, Doubleday, 1966.

Smith, Alice Lorraine. Microbiology laboratory manual and workbook. 2nd ed. St. Louis, Mo., Mosby, 1970.

Sober, Herbert A., ed. CRC handbook of biochemistry, selected data for molecular biology. 2nd ed. Cleveland, Chemical Rubber Co., 1970.

Stebbins, Robert Cyril. A Field guide to western reptiles and amphibians: field marks of all species in western North America. Boston, Houghton-Mifflin, 1966.

Sumner, Adrian Thomas. A Laboratory manual of microtechnique and histochemistry. Oxford, Blackwell Scientific, 1969.

Thompson, Clem W. Manual of structural kinesiology. 6th ed. St. Louis, Mosby, 1969.

U. F. A. W. Handbook of the care & management of laboratory animals. 3rd ed. Baltimore, Williams & Wilkins, 1967.

Viktorov, Sergei V., et al. Short guide to geo-botanical surveying. Elmsford, N. Y., Pergamon, 1964.

Vincent, J. M. A Manual for the practical study of root-nodule bacteria. Philadelphia, F. A. Davis, 1970.

Walden, H. T. Familiar fresh water fishes of America. New York, Harper & Row, 1964.

Walker, Ernest P., et al. Mammals of the world. 2nd ed. Baltimore, Johns Hopkins Press, 1968. 3 v.

Wang, Yen, ed. CRC handbook of radioactive nuclides. Cleveland, Chemical Rubber Co., 1969.

Wayne, Philip. A Guide to the pheasants of the world. New York, Country Life Bks., 1969.

Webb, Walter F. Handbook for shell collectors. 16th rev. ed. Wellesley Hills, Mass., Lee Pubns., 1966.

Wetmore, Alexander, et al. Song and garden birds of North America. With records. Washington, D. C., National Geo. Sci., 1964.

__________. Water, prey, & game birds of North America. With records. Washington, D. C., National Geo. Sci., 1965.

Wharton, Mary E. A Guide to the wildflowers and ferns of Kentucky. Lexington, Ky., Univ. Pr. of Kentucky, 1971.

Williams, John George. A Field guide to the butterflies of Africa. Boston, Houghton-Mifflin, 1971.

Winchester, A. M. Concepts of zoology: laboratory manual. Dubuque, Iowa, Wm. C. Brown, 1970.

Yudkin, M. & Offord, R. Harrison's guidebook to biochemistry. 3rd rev. ed. New York, Cambridge Univ. Press, 1971.

MEDICAL SCIENCES

Ackner, Brian, ed. Handbook for psychiatric nurses. 9th ed. Baltimore, Williams & Wilkins, 1964.

Air pollution control; guidebook for management. Stamford, Conn., Environmental Science Service Div., E. R. A. Inc., 1969.

American Association for Health, Physical Education and Recreation. Drug resource book for drug abuse education. Washington, D. C., A. A. H. P. E. R., 1970.

American Pharmaceutical Association. National formulary. 13th ed. Washington, D. C., Mack Pub., 1970.

American Physiological Society. Handbook of physiology. Baltimore, Williams & Wilkins, v. 1, 1959--

American Public Health Association. Program Area Committee on Air Pollution. Guide to the appraisal and control of air pollution. 2nd ed. New York, American Public Health Ass'n., 1969.

Annan, Gertrude L. & Felter, Jaquiline W., eds. Handbook of medical library practice. 3rd ed. Chicago, Med. Lib. Assn., 1970.

Armengal, Joseph, and others, comp. English-Spanish guide for medical personnel. Flushing, N. Y., Medical Exam., 1966.

Baillière's handbook of first aid: an elementary and advanced course of training. 6th ed. Baillière, Tindall & Cassell, dist. by Baltimore, Williams & Wilkins, 1970.

Bauer, William Waldo. Today's health guide; a manual of health information & guidance for the American family. 2nd rev. ed. Chicago, American Medical Association, 1970.

Belinkoff, Stanton. Manual for the recovery room. Boston, Little, 1967.

Beneke, Everett Smith & Rogers, Alvin Lee. Medical mycology manual. 3rd ed. Minneapolis, Burgess, 1971.

Benson, Harold J. Anatomy and physiology: laboratory. Dubuque, Iowa, Wm. C. Brown, 1970.

Benson, Ralph C. Handbook of obstetrics & gynecology. 3rd ed. Los Altos, Cal., Lange, 1968.

Betschman, Lucille I. Handbook of recovery room nursing. Philadelphia, Davis Co., 1967.

Better Homes and Gardens, ed. Family medical guide. New York, Meredith, 1964.

Bleier, Inge J. Workbook in bedside maternity nursing. Philadelphia, Saunders, 1969.

Blinkov, Samuel M. & Glezer, Ilya I. The Human brain in figures and tables: a quantitative handbook. New York, Basic Books, 1968. (Tr. from the 1964 Russian edition.)

Boutkan, J. ABC of the EGG, a guide to electrocardiography. Springfield, Ill., C. C. Thomas, 1970, c1969.

Boyd, W. C. Pathology for the physician. 8th ed. Philadelphia, Lea & Febiger, 1967.

Bray, William E. Bray's clinical laboratory methods. Ed. by Bauer, John Dictal. 7th ed. St. Louis, Mosby, 1968.

Bredow, Miriam. Medical secretarial procedures. 5th ed. New York, McGraw-Hill, 1966.

__________ & Cooper, Marian G. The Medical assistant; a guide to clinical, secretarial, and technical duties. 3rd ed. New York, McGraw-Hill, 1970.

Brown, Warren J. Patients' guide to medicine; from the drugstore through the hospital. Largo, Fla., Aero-Medical Consultants, Inc., 1969.

Burack, Richard. The New handbook of prescription drugs: official names, prices, and sources for patient and doctor. New York, Ballantine, 1970, c1967.

Burgess, Audrey. The Nurse's guide to fluid and electrolyte balance. New York, McGraw-Hill, 1970.

Chatton, Milton J., et al. Handbook of medical treatment. 12th ed. Los Altos, Cal., Lange, 1970.

Chernok, Norma B., comp. Radiology typists handbook; a handy reference guide of medical terms used in radiologic reports for typists. Flushing, N.Y., Med. Exam. Pub., 1970.

Cohn, Helen, et al. Manual for nurses in family and community health. Boston, Little, 1969.

Cooper, Signe S. Contemporary nursing practice; a guide for the returning nurse. New York, McGraw-Hill, 1970.

Crawford, Annie Laurie & Buchanan, Barbara Boring. Psychiatric nursing; a basic manual. 3rd ed. Philadelphia, F. A. Davis, 1970.

Cunningham, D. J. Manual of practical anatomy. 13th ed. New York, Oxford Univ. Pr., 1968-1969. 3 v.

Current drug handbook. Comp. by Mary W. Falconer, et al. Philadelphia, Saunders, 1970.

DeKornfeld, Thomas J. & Gilbert, Don E. Inhalation therapy procedure manual. 2nd ed. Springfield, Ill., C. C. Thomas, 1970.

Delaney, John William & Garratty, George. Handbook of haematological and blood transfusion technique. 2nd ed. New York, Appleton, 1969.

DiMascio, Alberto & Shader, Richard I., eds. Clinical handbook of psychopharmacology. New York, Science House, 1970.

Dreisbach, Robert H. Handbook of poisoning: diagnosis & treatment. 6th ed. Los Altos, Cal., Lange, 1969.

Fishbein, Morris, ed. Modern home medical advisor: your health and how to preserve it. New York, Doubleday, 1969.

Garb, Solomon. Clinical guide to undesirable drug interactions and interferences. New York, Springer, 1971.

Geitgey, Doris Arlene. A Handbook for head nurses. 2nd

ed. Philadelphia, F. A. Davis, 1971.

Gleason, Marion N., et al. Clinical toxicology of commercial products; acute poisoning (home & farm). 3rd ed. Baltimore, Williams & Wilkins, 1969.

Grant, Murray. Handbook of preventive medicine & public health. Philadelphia, Lea & Febiger, 1967.

Gray, Henry. Gray's anatomy of the human body. Ed. by Goss, C. M. Philadelphia, Lea & Febiger, 1966.

Haines, R. Wheeler & Mohiuddin, A. Handbook of human embryology. 4th ed. Baltimore, Williams & Wilkins, 1968.

Hanok, Albert. Manual for laboratory clinical chemistry. Los Altos, Cal., Geron-X, 1969.

Henderson, John. Emergency medical guide. 2nd ed. New York, McGraw-Hill, 1969.

Hirschhorn, Howard H. Spanish-English, English-Spanish medical guide. New York, Regents Pub. Co., 1969.

Hirschhorn, Richard C. Handbook of practical urology. Philadelphia, Lea & Febiger, 1965.

Holm, Niels W. & Berry, Roger J., eds. Manual on radiation dosimetry. New York, M. Dekker, 1970.

Homan, William E. Child sense; a pediatrician's guide for today's families. New York, Bantam, 1970, c1969.

Horrobin, David & Gunn, Alexander. The International handbook of medical science; a concise guide to current practice and recent advances. New York, Putnam, 1970.

Hunter, George W., et al. Manual of tropical medicine. 4th ed. Philadelphia, Saunders, 1966.

International Planned Parenthood Federation. Handbook of oral contraception. Baltimore, Williams & Wilkins, 1965.

Joel, Alma L. Workbook and study guide for medical-surgical nursing; a patient-centered approach. 2nd ed. St.

Louis, Mosby, 1969, c1965.

Karelitz, Samuel. When your child is ill. Rev. ed. New York, Random House, 1969.

Kaye, Sidney. Handbook of emergency toxicology. 3rd ed. Springfield, Ill., C. C. Thomas, 1970. (American Lectures on Public Protection Series.)

Kodet, E. Russel & Angier, Bradford. The Home medical handbook. New York, Association Pr., 1970.

Krusen, Frank H., et al, eds. Handbook of physical medicine & rehabilitation. Philadelphia, Saunders, 1965.

Leake, Mary J. Manual of simple nursing procedures. 4th ed. Philadelphia, Saunders, 1966.

Lerch, Constance. Workbook for maternity nursing. 2nd ed. St. Louis, Mosby, 1970.

Lillywhite, Herold S. Pediatricians handbook of communication disorders. Philadelphia, Lea & Febiger, 1970.

Mayo Clinic. Rochester, Minn. Committee on Dietetics. Mayo Clinic diet manual. 4th ed. Philadelphia, Saunders, 1971.

Meering, A. B. Handbook for nursery nurses. 4th ed. Baltimore, Williams & Wilkins, 1964.

Merck manual of diagnosis & therapy. 11th ed. Rahway, N. J., Merck, 1966.

Miller, Benjamin F. and Galton, Lawrence. The Family book of preventive medicine: how to stay well all the time. New York, Simon & Schuster, 1971.

Mitchell, John P. & Lumb, Geoffrey N. Handbook of surgical diathermy. Baltimore, Williams & Wilkins, 1966.

Modell, Walter, et al. Handbook of cardiology for nurses. 5th ed. New York, Springer Pub., 1966.

Morris, Stephen Cripwell. A Complete handbook for professional ambulance personnel. Baltimore, Md., Williams & Wilkins, 1970.

Physician's handbook. 16th ed. Los Altos, Cal., Lange, 1970.

Price, Alice L. Handbook and charting manual for student nurses. 4th ed. St. Louis, Mosby, 1967.

Pryor, William J. Manual of anesthetic techniques. 3rd ed. Baltimore, Williams & Wilkins, 1966.

Ralston, Edgar L. Handbook of fractures. St. Louis, Mosby, 1967.

Rossman, I. J. & Schwartz, D. R. The Family handbook of home nursing and medical care. Philadelphia, Lippincott, 1968.

Sauer, Gordon C. Manual of skin diseases. 2nd ed. Philadelphia, Lippincott, 1966.

Shepard, Kenneth S. Care of the well baby: medical management of the child from birth to 2 years of age. Philadelphia, Lippincott, 1968.

Shestack, Robert. Handbook of physical therapy. 2nd ed. New York, Springer, 1967.

Silver, Henry K., et al. Handbook of pediatrics. 9th ed. Los Altos, Cal., Lange, 1971.

Sobol, Evelyn G. & Robischon, Paulette. Family nursing; a study guide. St. Louis, Mosby, 1970.

Stiles, Karl A. Handbook of histology. 5th ed. New York, McGraw-Hill, 1968.

Sutton, Audrey Latshaw. Workbook for practical nurses. 3rd ed. Philadelphia, Saunders, 1970.

U.S. Dispensatory and Physicians Pharmacology. 26th ed. Philadelphia, Lippincott, 1967. (title varies.)

Vinken, P. J. & Bruyn, G. W. Handbook of clinical neurology. New York, American Elsevier, 1968-1970. 13 v.

Weir, D. M. Handbook of experimental immunology. Philadelphia, Davis Co. (Blackwell), 1967.

Wilson, J. L., et al. Handbook of surgery. 4th ed. Los Altos, Cal., Lange, 1969.

Zimmerman, Clarence E. Techniques of patient care; a manual of bedside procedures for students, interns and residents. Boston, Little, Brown & Little, 1970.

AGRICULTURAL SCIENCES

Abraham, G. The Green thumb of indoor gardening; a complete guide. New York, Prentice-Hall, 1967.

American Kennel Club. The Complete dog book. New York, Doubleday, 1969.

Barnes, C. D. & Eltherington, L. G. Drug dosage in laboratory animals; a handbook. Berkeley, Cal., Univ. of Cal. Pr., 1964.

Cone, Arthur L., Jr. The Complete guide to hunting. New York, Macmillan, 1970.

Ensminger, M. Eugene. The Stockman's handbook. 4th ed. Danville, Ill., Interstate, 1970.

Farm chemical handbook. Willoughby, Ohio, Meister Pub. Co., 1st ed., 1908-- (title varies.)

Gilbert, John. The Complete aquarist's guide to freshwater fishes. New York, Golden Press, 1970.

Graf, Alfred Byrd. Exotic plant manual; fascinating plants to live with, their requirements and use. East Rutherford, N.J., Roehrs, 1970.

Guthrie, E. L. & Miller, R. C. Home book of animal care. New York, Harper & Row, 1966.

Higham, Robert R. Handbook of papermaking: the technology of pulp, paper, & board manufacture. New York, International Pubn. Serv., 1968.

Hodge, Peggy Hickok. Tropical gardening; handbook for the home gardener. Rutland, Vt., Tuttle, 1970.

Jennings, A. R. Animal pathology: a concise guide to

systematic veterinary pathology for students and practitioners. London, Baillière, Tindall & Cassell, 1970.

Kirk, Robert W. & Bistner, Stephen I. Handbook of veterinary procedures & emergency treatment. Philadelphia, Saunders, 1969.

Merck veterinary manual. 3rd ed. Rahway, N. J., Merck, 1967.

Midwest farm handbook. Ames, Iowa, Iowa State Univ. Pr., 1st ed., 1949--

Morris, Dan & Strung, Norman. The Fisherman's almanac. New York, Macmillan, 1970.

Pesticide handbook -- entoma 1968. 20th ed., State College, Pa., College Science Pub., 1968.

Snyder, Rachel, ed. The Complete book for gardeners. New York, Van Nostrand-Reinhold, 1964.

Sprague, Howard Bennett. Turf management handbook. Danville, Ill., Interstate, 1970.

Taylor, Norman, ed. The Guide to garden shrubs and trees. Boston, Houghton-Mifflin, 1965.

U. S. Department of Agriculture. Handbook of agricultural charts. Washington, D. C., U. S. G. P. O., 1963--

__________. Suggested guide for weed control. Washington, D. C., 1969. (Agriculture Handbook No. 332.)

__________. Forest Service. Silvics of forest trees of the U. S. Washington, D. C., 1965. (Agriculture Handbook No. 271.)

Westcott, Cynthia. Plant disease handbook. 3rd ed. New York, Van Nostrand-Reinhold, 1971.

Wolfenstine, Manfred R. The Manual of brands and marks. Norman, Okla., Univ. of Okla. Pr., 1970.

ENGINEERING SCIENCES

AGARD. Manual on aircraft loads. Ed. by Taylor, J. Elmsford, N.Y., Pergamon, 1965.

Abbey, Stanton. Book of the Volkswagen: all models up to 1968. 4th ed. New Rochelle, N.Y., Soccer, 1969. (Pitman's Motorists' Library Series.)

Albers, Vernon M. Underwater acoustics handbook. 2nd ed. University Park, Pa., Pennsylvania State Univ. Pr., 1965.

American Gas Association. Gas engineers handbook; fuel gas engineering practice. Ed. by Seigeler, C. George. New York, Industrial Press, 1965.

American Home Economics Association. Textile handbook. 4th ed. Washington, D.C., 1966.

American Society for Metals. Metals handbook. Metals Park, Ohio, 1st ed., 1927--. (5 v., 8th ed. 1961--.)

American Society for Testing and Materials. Book of ASTM standards. Philadelphia, ASTM, 1st ed., 1910--. 33 parts. (8th ed., 1970.)

American Society of Mechanical Engineers. ASME handbook: metals engineering-design. Ed. by Horger, O. J. 2nd ed. New York, McGraw-Hill, 1965.

American Society of Tool and Manufacturing Engineers. ASTME die design handbook. Ed. by Wilson, Frank W. 2nd ed. New York, McGraw-Hill, 1965.

__________. ASTME handbook of industrial metrology. New York, Prentice-Hall, 1967.

American Water Works Association. Water quality and treatment manual. 3rd ed. New York, McGraw-Hill, 1970.

Amrine, Michael. The Safe driving handbook. New York, Grosset & Dunlap, 1970.

Appels, J. R. & Geels, B. H. Handbook of relay switching technique. New York, Springer-Verlag, 1966.

Auerbach Info. Inc. Auerbach guide to data communications. Philadelphia, Auerbach Pubs., 1970.

Babcoke, Carl H. RCA monochrome TV service manual. Blue Ridge Summit, Pa., G/L Tab Books, 1970.

Bakish, Robert & White, S. S. Handbook of electron beam welding. New York, Wiley, 1964.

Barton, David Knox & Ward, Harold R. Handbook of radar measurement. New York, Prentice-Hall, 1969.

Baumeister, T. & Marks, Lionel. Standard handbook for mechanical engineers. 7th ed. New York, McGraw-Hill, 1967.

Belt, Forest H. Motorola color TV service manual. Blue Ridge Summit, Pa., Tab Books, 1969.

Benedict, Robert P. & Carlucci, Nicola A. Handbook of specific losses in flow systems. New York, Plenum (IFI), 1966.

__________ & Steltz, W. G. Handbook of generalized gas dynamics. New York, Plenum (IFI), 1966.

Bland, William F. & Davidson, R. L. Petroleum processing handbook. New York, McGraw-Hill, 1967.

Bolz, Ray E., ed. CRC handbook of tables for applied engineering science. Cleveland, Chemical Rubber Co., 1970.

Boyce, William F. Hi-fi stereo handbook. 3rd ed. Indianapolis, Ind., Sams, 1967.

Bradley, John H. Programmer's guide to the IBM system/360. New York, McGraw-Hill, 1969.

Brady, George Stuart. Materials handbook; an encyclopedia for purchasing managers, engineers, executives and foremen. 10th ed. New York, McGraw-Hill, 1971.

British Interplanetary Society. Handbook of astronautics. Ed. by Smith, S. W. Chester Springs, Pa., Dufour, 1969.

Britt, Kenneth W., ed. Handbook of pulp and paper technology. 2nd ed. rev. & enl. New York, Van Nostrand-Reinhold, 1970.

Brodbeck, Emil E. Handbook of basic motion picture techniques. Philadelphia, Chilton, 1966.

CRC Fenaroli's handbook of flavor ingredients. International ed. Cleveland, Chemical Rubber Co., 1971.

Cagle, Charles V. Adhesive bonding; techniques and applications. New York, McGraw-Hill, 1968.

Carrier Air Conditioning Co. Handbook of air conditioning system design. New York, McGraw-Hill, 1966.

Chilton's auto repair manual. Philadelphia, Chilton, 1970.

Christensen, James J. Handbook of metal liquid heats and related thermodynamic quantities. New York, M. Dekker, 1970.

Clauss, Francis Jacob. Engineer's guide to high-temperature materials. Reading, Mass., Addison-Wesley, 1969.

Clifford, Martin. Electronics data handbook. New York, Tab Books, 1964.

Coker, A. J. Newnes motor repair. New York, Arco, 1966. 5 v. and 24 wall charts.

Collier, Ann M. A Handbook of textiles. Elmsford, N. Y., Pergamon, 1971.

Collins, Archie Frederick. The Radio amateur's handbook. 12th ed. New York, T. Y. Crowell, 1970.

Compressed Gas Association. Handbook of compressed gases. New York, Van Nostrand-Reinhold, 1966.

Considine, Douglas M. & Ross, S. D., eds. Handbook of applied instrumentation. New York, McGraw-Hill, 1964.

Croft, Terrell, Carr, Clifford C. & Watt, John H. American electrician's handbook. 9th ed. New York, McGraw-Hill, 1970.

Crosby, Edward G. and Kochis, Stephen N. Plastic applications handbook. Boston, Cahners, 1971.

Crowhurst, N. H. Electronics reference databook. Blue Ridge Summit, Pa., TAB Books, 1969.

Daellenbach, Hans G. User's guide to linear programming. New York, Prentice-Hall, 1970.

Davidson, A. Handbook of precision engineering. v. 1: fundamentals, v. 2: materials. New York, McGraw-Hill, 1971, c1970.

Davis, Calvin V. & Sorenson, K. E. Handbook of applied hydraulics. 3rd ed. New York, McGraw-Hill, 1969.

Day, Richard. The Practical handbook of concrete and masonry. New York, Arco, 1969.

__________. The Practical handbook of electrical repairs. New York, Arco, 1969.

__________. The Practical handbook of plumbing and heating. New York, Arco, 1969.

DeCristoforo, R. J. The Practical handbook of carpentry. New York, Arco, 1969.

Deschin, Jacob. New York Times guide to taking better pictures. New York, Golden Press, 1969.

Eimbinder, Jerry. FET applications handbook. 2nd ed. (Field Effect Transistors.) Blue Ridge Summit, Pa., Tab Books. 1970.

Electronic technician TV troubleshooter's handbook. 2nd ed. Electronic Technician/Dealer, eds. Blue Ridge Summit, Pa., Tab Books, 1970.

Engelman, Roy Albert. Engleman's autocraft. Philadelphia, Chilton, 1970.

Feinberg, R., ed. Handbook of electronic circuits. New York, Barnes & Noble, 1966.

Feldzamen, Alvin N. The Intelligent man's easy guide to computers. New York, McKay, 1971.

Fink, Donald G. & Carroll, John M., eds. Standard handbook for electrical engineers. 10th ed. New York, McGraw-Hill, 1968.

Fomenko, Vadim S. & Samsonov, G. W. Handbook of thermionic properties. New York, Plenum (IFI), 1966.

Francis, Mary. The Beginner's guide to flying. London, Pelham Books, 1969.

Furia, Thomas E. CRC Handbook of food additives. Cleveland, Chemical Rubber Co., 1968.

Gartman, Hans. DeLaval engineering handbook. 3rd ed. New York, McGraw-Hill, 1970.

Gass, Saul I. An Illustrated guide to linear programming. New York, McGraw-Hill, 1970.

Gaylord, Edwin H. & Charles N. Structural engineering handbook. New York, McGraw-Hill, 1968.

Giebelhausen, J. Manual of applied photography. (English-German). New York, Museum Books, 1966.

Glenn's foreign car repair manual. 2nd ed., Philadelphia, Chilton, 1966.

Glushkov, G. S., et al. Handbook of formulas for the analysis of complex frames & arches. Hartford, Conn., Davey, 1967.

Goodman, Robert L. Philco color TV service manual. Blue Ridge Summit, Pa., Tab Books, 1970.

Graf, Rudolf F. Electronic design data book. New York, Van Nostrand-Reinhold, 1971.

Griffel, William. Handbook of formulas for stress & strain. New York, Ungar, 1966.

Griffin, Al. So you'd like to buy an airplane; a handbook for prospective owners. New York, Macmillan, 1970.

Griffin, Charles William for the American Institute of Architects. Manual of built-up roof systems. New York, McGraw-Hill, 1970.

Gross, William F. Applications manual for paint and protective coatings. New York, McGraw-Hill, 1970.

Gruenberg, E. L. Handbook of telemetry & remote control. New York, McGraw-Hill, 1967.

Guilbault, George C. Instrumental analysis manual; modern experiments for the laboratory. New York, M. Dekker, 1970.

Hall, Archibald J. Handbook of textile finishing. Rev. ed. Metuchen, N. J., Textile Book, 1966.

__________. Standard handbook of textile. New York, Chemical Pub., 1970.

Handbook of fiberglass and advanced plastics composites. Ed. by Lubin, George. New York, Van Nostrand-Reinhold, 1969.

Handbook of industrial research management. 2nd ed., New York, Van Nostrand-Reinhold, 1968.

Handley, William. Industrial safety handbook. New York, McGraw-Hill, 1970.

Hanlon, Joseph F. Package engineering handbook. New York, McGraw-Hill, 1971.

Harper, Charles A. Handbook of electronic packaging. New York, McGraw-Hill, 1969.

__________. Handbook of materials and processes for electronics. New York, McGraw-Hill, 1970.

Harrington, Donald E. & Meacham, Stanley. Handbook of electronic tables & formulas. 3rd ed. Indianapolis, Ind., Sams, 1968.

Havers, John A. Handbook of heavy construction. 2nd ed. New York, McGraw-Hill, 1971.

Haviland, R. P. Handbook of satellites & space vehicles. New York, Van Nostrand-Reinhold, 1965.

Hawkins, George A., ed. Student's engineering manual. New York, McGraw-Hill, 1968.

Hickin, N. E. Wood preservation: a guide to the meaning of terms. London, British Book Centre, 1971.

Hicks, David E. Citizens band radio handbook. 3rd ed., Indianapolis, Ind., Sams, 1967. (Title varies.)

Higham, Robert R. A. A Handbook of paperboard and board; its manufacturing technology, conversion and usage. London, Business Books, 1970.

Hughes, L. E. C. and Holland, F. W., eds. Handbook of electronic engineering. 3rd ed. Cleveland, Chemical Rubber Co., 1967.

Hunter, Lloyd P. Handbook of semiconductor electronics. 3rd ed., New York, McGraw-Hill, 1970.

International Atomic Energy Agency. Manual of safety aspects of the design and equipment of hot laboratories. New York, 1969.

Johnson, Francis S., ed. Satellite environment handbook. 2nd ed., Stanford, Cal., Stanford Univ. Pr., 1965.

Kershner, William K. The Advanced pilot's flight manual. 3rd ed., Ames, Iowa, Iowa State Univ. Press, 1970.

Keys, William J. & Powell, Carl H. A Handbook of modern keypunch operations. San Francisco, Canfield Pr., 1969, c1970.

King, Gordon John. The Hi-fi and tape recorder handbook. Levittown, N.Y., Transatlantic, 1971.

Kinori, B. Z. Manual of surface drainage engineering. New York, American Elsevier, 1970.

Klapper, Marvin. Fabric almanac. New York, Fairchild Pubns., 1966--

Kohl, Walter H. Handbook of materials & techniques for vacuum devices. New York, Van Nostrand-Reinhold, 1967.

Kozlov, Boris Anatol'evich & Ushakov, I. A. Reliability handbook. New York, Holt, Rinehart & Winston, c1970.

Lenk, John D. Applications handbook for electrical connectors. Indianapolis, Ind., Sams, 1966.

_________. Data book for electronic technicians & engineers. New York, Prentice-Hall, 1968.

_________. Handbook of electronic charts, graphs, and tables. New York, Prentice Hall, 1970.

_________. Handbook of electronic meters: theory & application. New York, Prentice-Hall, 1969.

_________. Handbook of electronic test equipment. New York, Prentice-Hall, 1971.

_________. Handbook of oscilloscopes: theory & application. New York, Prentice-Hall, 1968.

_________. Handbook of practical electronic tests and measurements. New York, Prentice-Hall, 1969.

_________. Handbook of simplified solid state circuit design. New York, Prentice-Hall, 1971.

_________. Practical semiconductor data book for electronic engineers & technicians. New York, Prentice-Hall, 1970.

Leslie, William Henderson Paterson. Numerical control users: handbook. New York, McGraw-Hill, 1970.

Liebers, Arthur. The Engineer's handbook. New York, Key, 1968.

Lockheed Missiles and Space Company. Space materials handbook. Reading, Mass., Addison-Wesley, 1966.

Lund, Herbert F., ed. Industrial pollution control handbook. New York, McGraw-Hill, 1971.

Lunts, G. L., et al, eds. Handbook for engineers. v. 1, mathematics and physics; v. 2, mechanics, strength of materials and the theory of mechanisms and machines. Elmsford, N.Y., Pergamon, 1964-1967. 2 v.

McClain, Thomas B. Complete handbook on environmental control; a reference manual for debaters and others interested in the subject. Skokie, Ill., National Textbook, 1970.

McEntee, Howard G. & Winter, William. The Model aircraft handbook. 5th ed. New York, T. Y. Crowell, 1968.

Machol, Robert E., et al. System engineering handbook. New York, McGraw-Hill, 1965.

McLaughlin, Terence. The Cleaning, hygiene and maintenance handbook. London, Business Books, 1969.

Maissel, Leon I. & Reinhard, Glang, eds. Handbook of thin film technology. New York, McGraw-Hill, 1970.

Mangulis, Visvaldis. Handbook of series for scientists & engineers. New York, Academic Press, 1965.

Manning, G. E. Weather radar for pilots: a handbook. New York, British Info, 1970.

Margolis, Art. The Practical handbook of TV repairs. New York, Arco, 1969.

Martin, Janette C. Manual of applied nutrition. 5th ed. Baltimore, Johns Hopkins, 1966.

Mechanical press handbook. Ed. by Daniels, Harold R. 3rd rev. & enl. ed. Boston, Cahners Pub., 1969.

Merritt, Frederick S. Building construction handbook. 2nd ed. New York, McGraw-Hill, 1965.

__________. Standard handbook for civil engineers. New York, McGraw-Hill, 1968.

Middleton, Robert Gordon. Hi-fi stereo servicing guide. Indianapolis, Ind., Sams, 1970.

__________. Radio receiver servicing guide. Indianapolis, Ind., Sams, 1970.

__________. Tape recorder servicing guide. Indianapolis, Ind., Sams, 1970.

National Research Council, Canada. Asso. Committee on Geotechnical Research. Muskeg Subcommittee. Muskeg engineering handbook. Toronto, Univ. of Toronto Pr., 1969.

Noddings, Charles R. & Mullet, G. M. Handbook of compositions at thermodynamic equilibrium. New York, Wiley (Interscience), 1965.

Noemer, Ewald F. Handbook of modern halftone photography with complete concepts & practices. Rev. ed. Demarest, N. J., Perfect Graphic Arts, 1970.

Noll, Edward M. First-class radio telephone license handbook. 3rd ed. Indianapolis, Ind., Sams, 1970.

Oberg, Erik & Jones, Franklin D. Machinery's handbook. Ed. by Horton, Holbrook L. 19th ed. New York, Industrial Press, 1971.

__________. Use of handbook tables & formulas for Machinery's handbook. Ed. by Horton, Holbrook L. 18th ed. New York, Industrial Press, 1968.

O'Conner, J. J. & Boyd, J., eds. Standard handbook of lubrication engineering. New York, McGraw-Hill, 1968.

Oleesky, Samuel S. & Mohr, J. Gilbert. Handbook of reinforced plastics of the Society of Plastics Industry. New York, Van Nostrand-Reinhold, 1964.

Ollard, Eric A. & Smith, E. B. Handbook of industrial electroplating. 3rd ed. New York, American Elsevier, 1964.

Palmquist, Roland E. Guide to the 1968 National electrical code. 2nd ed. Indianapolis, Ind., Audel, 1969.

Parker, Earl R. Materials data book. New York, McGraw-Hill, 1967.

Pearson, W. B. A Handbook of lattice spacings and structures of metals and alloys. Elmsford, N. Y., Pergamon, 1958-1967. 2 v.

Perry, Robert H. Engineering manual; a practical reference of data and methods in architectural, chemical, civil, electrical, mechanical, and nuclear engineering. 2nd ed., New York, McGraw-Hill, 1967.

Plunkett, Edmond R. Handbook of industrial toxicology. New York, Chem. Pub., 1966.

Plutonium handbook; a guide to the technology. Ed. by O. J. Wick. New York, Gordon & Breach, 1967. 2 v.

Poster, Arnold R. Handbook of metal powders. New York, Van Nostrand-Reinhold, 1966.

Purdue University Thermophysical Properties Research Centers. Handbook of thermophysical properties of high temperature solid materials. Ed. by Touloukion, T. S. New York, Macmillan, 1967. 6 v.

Pyle, H. S. General class amateur license handbook. 2nd ed. Indianapolis, Ind., Sams, 1968. (title varies.)

Radar cross section handbook. Ed. by George T. Ruck. New York, Plenum, 1970. 2 v.

Radio amateur's handbook. Rev. annually. Newington, Conn., American Radio Relay League, 1926--

Radio handbook. 18th ed., Summerland, Cal., Editors & Engineers, 1968.

Reed, George Henry. Refrigeration: a practical manual for apprentices. New York, Hart, 1970, c1967.

Rider, J. F. Perpetual troubleshooter's manual. New York, v. 1, 1933-- (looseleaf.)

__________. Television manual. New York, v. 1, 1948-- (looseleaf.)

Rohsenow, Warren and Hartnett, J. P. Handbook of heat transfer. New York, McGraw-Hill, 1971.

Rossnagel, W. E. Handbook of rigging in construction & industrial operations. 3rd ed. New York, McGraw-Hill, 1964.

Russell, P. J. Sea signalling simplified: a manual of instruction for the new International Code of Signals. 3rd ed. London, Coles, 1969.

Safford, E. L., Jr. A Guide to radio & TV broadcast engineering practice. Blue Ridge Summit, Pa., Tab Books, 1971.

Sams (Howard W.) and Company Editorial Staff. Color-TV training manual. 3rd ed. Indianapolis, Ind., Sams, 1970.

Sams (Howard W.) and Company, Inc. Transistor specifications manual. 4th ed. Indianapolis, Ind., Sams, 1970.

__________. Transistor substitution handbook. 10th ed. Indianapolis, Ind., Sams, 1970.

__________. Tube substitution handbook. 13th ed. Indianapolis, Ind., Sams, 1970.

Sands, Leo G. CB radio servicing guide. 2nd ed. Indianapolis, Ind., Sams, 1969.

Schultz, John J. Electronic test & measurement handbook. Blue Ridge Summit, Pa., Tab Books, 1969.

Schultz, Morton. The Practical handbook of painting and wallpapering. New York, Arco, 1969.

Schweitzer, Philip A. Handbook of corrosion resistant piping. New York, Industrial Press, 1969.

Shane, Jay. Home-call TV repair guide. Blue Ridge Summit, Pa., Tab Books, 1970.

Shea, Richard F. Amplifier handbook. New York, McGraw-Hill, 1966.

Shekter, Robert J. Standard handbook of pleasure boats. New York, T. Y. Crowell, 1970, c1959.

Shields, J. Adhesives handbook. Cleveland, Chemical Rubber Co., 1971.

Simonson, Leroy. Private pilot study guide. Glendale, Cal., Aviation Book Co., 1970.

Skolnik, Merrill Ivan, ed. Radar handbook. New York, McGraw-Hill, 1970.

Smeaton, Robert W., ed. Motor application and maintenance handbook. New York, McGraw-Hill, 1969.

Smothers, William J. & Chiang, Yao. Handbook of differential thermal analysis. Rev. ed. New York, Chem. Pub., 1966.

Society of Automotive Engineers. S.A.E. handbook. New York, 1924-- (Annual.)

Souders, Mott. The Engineer's companion; a concise handbook of engineering fundamentals. New York, Wiley, 1966.

Squires, Terence Leighton, ed. Telecommunications pocket book. Levittown, N.Y., Transatlantic, 1971.

Stetka, Frank. NFPA handbook of the national electrical code. 2nd ed. New York, McGraw-Hill, 1969.

Steward, Robert M. Boatbuilding manual. New York, Taplinger, 1969.

Strock, Clifford & Koral, Richard L., eds. Handbook of air conditioning, heating, & ventilating. 2nd ed. New York, Industrial Press, 1965.

Swatek, Paul. The User's guide to the protection of the environment. New York, Friends of the Earth/Ballantine, 1970.

TAB Editorial Staff, ed. Popular tube/transistor substitution guide. 2nd ed. Blue Ridge Summit, Pa., Tab Books, 1970.

Thomas, H. Handbook of transistors, semiconductors, instruments & microelectronics. New York, Prentice-Hall, 1969.

Thomas, Harry Elliot. Handbook for electronics engineers and technicians. New York, Prentice-Hall, 1965.

__________. Handbook of integrated circuits. New York, Prentice-Hall, 1971.

__________. Handbook of pulse digital devices for communication data processing. New York, Prentice-Hall, 1970.

__________ & Clarke, C. A. Handbook of electronic instruments and measurement techniques. New York, Prentice-Hall, 1967.

Thornton, Richard D., et al. Handbook of basic transistor circuits and measurements. New York, Wiley, 1966.

Tsiklis, D. S. Handbook of techniques in high pressure research & engineering. Ed. by Bobrowsky, Alfred. New York, Plenum Press, 1968.

Ulrey, Harry F. Building construction and design. 2nd ed. Indianapolis, Ind., Audel, 1970.

U.S. Aerospace Studies Institute. Communications-electronics terminology handbook. Washington, D.C., Public Affairs Press, 1965.

U.S. Atomic Energy Commission. Reactor handbook. 2nd ed., New York, Wiley (Interscience), 1964. 4 v.

U.S. Federal Aviation Administration. Flight Information Division. Airman's information manual. Ed. by Walter P. Winner. Glendale, Cal., Aviation Book Co., 1970.

Von Aulock, Wilhelm H. Handbook of microwave ferrite materials. New York, Academic Press, 1965.

Walker, William Francis. Beginner's guide to jig and tool design. New York, Hart, 1969, c1967.

Walraven, H. Dale. Handbook of engineering graphics. New York, McKnight, 1965.

Walton, J. D., Jr. Radome engineering handbook; design and principles. New York, M. Dekker, 1970.

Wass, Alonzo. Manual of structural details for building construction. New York, Prentice-Hall, 1968.

Westman, H. P., ed. Reference data for radio engineers. 5th ed., Indianapolis, Ind., Sams, 1968.

White, Robert G. Handbook of industrial infrared analysis. New York, Plenum Press, 1964.

Woodruff, H. Charles. Short-wave listener's guide. 4th

ed. Indianapolis, Ind., Sams, 1970.

Woolrich, Willis R. Handbook of refrigerating engineering. v. 1, fundamentals, 1965; v. 2, applications, 1966. Westport, Conn., Avi, 1965-1966. 2 v.

Wortman, Leon A. Closed-circuit television handbook. Indianapolis, Ind., Sams, 1970.

Zaba, Joseph and Doherty, W. T. Practical petroleum engineering handbook. 5th ed. Houston, Gulf Publishing Co., 1969.

Atlases

Geographical atlases are standard reference books usually found in most libraries. In cartography, an atlas is a collection of maps usually arranged alphabetically by region plus facts and figures about places in a bound volume. Atlases are divided into four classes: 1) general world, 2) special world, 3) general regional, and 4) special regional. They contain many kinds of maps: topographical, climatic, geologic, economic and political.

Indexes to atlases usually give coordinates for locating geographical features. The index may also include a gazetteer, or a listing of geographical names plus descriptions. The gazetteer, or dictionary of places, is helpful if the library search requires current information about a county, state, province, or nation: present population, importance, industries, and political affiliations. Only the most recent editions will serve since the older works may be almost worthless or even misleading.

Most historians credit Claudius Ptolemy, a geographer who lived in Egypt, with publishing the first atlas. It appeared in the A. D. 100's as part of an eight-volume work on map making.

The first modern general atlas, entitled Theatrum orbis terrarum, was published by Abraham Ortelius of Antwerp in 1570. The word "atlas" was used first by Gerardus Mercator in the title of his map collection Atlas sive cosmographical meditationes, 1585-1595, because a picture of the Greek mythological figure Atlas supporting the world on his shoulders had been used as an introductory illustration. Gradually the name came to be applied to volumes of maps or charts and sometimes to non-geographical publications of similar format.

Most general encyclopedias also include maps, either in one atlas volume or spread through the volumes illustrating the articles on the various regions and nations of the world.

The term "atlas" also describes a technical handbook which presents drawings, photographs, charts, plates, or tables illustrating one subject. These special atlases are useful when photos and illustrations are needed for graphic representation of a subject such as anatomy, astronomy or plant pathology.

Some of the more useful atlases are:

General Drafting Company, Inc. Man's domain; a thematic atlas of the world. New York, McGraw-Hill, 1969.

Goode, John Paul. World atlas. Ed. by Espenshade, Edward B. Jr. 13th ed. Chicago, Rand McNally, 1970.

Hammond Inc. Hammond contemporary world atlas. New York, Doubleday, 1969.

__________. Hammond standard world atlas. Maplewood, N.J., Hammond Inc., 1969.

__________. Headline world atlas. Maplewood, N.J., Hammond Inc., 1969.

The National atlas of the United States of America. Arlington, Va., Washington Distribution Section, U.S. Geological Survey, 1971.

Rand McNally and Company. The World book atlas. Chicago, Field Enterprises Educ. Corp., 1970.

Rand McNally Editors. Rand McNally new cosmopolitan world atlas. Rev. ed. Chicago, Rand McNally, 1968.

Shepherd, William R. Shepherd's historical atlas. 9th ed., New York, Barnes & Noble, 1964.

Times atlas of the world (English). London, London Times, 1956-1960. 5 v.

U.S. Copyright Office. Catalog of copyright entries: Third Series, Part 6, Maps and atlases. Washington, D.C., U.S.G.P.O., v. 1-- Jan./June 1947--

Walsh, S. Padraig, comp. Home reference books in print: atlases, English language dictionaries & subscription books. New York, R. R. Bowker, 1969.

ASTRONOMY

Alter, Dinsmore, ed. Lunar atlas. Magnolia, Mass., Peter Smith, 1968. Reprint of work originally published in 1964 by North American Aviation Inc. in limited edition.

California Institute of Technology. Jet Propulsion Laboratory. Ranger VII. Photographs of the moon. Washington, D.C., N.A.S.A., 1964-1965. 3 v.

DeCallatay, Vincent & Dollfus, Audouin. Atlas of the planets. Toronto, Univ. of Toronto Pr., 1970.

Kopal, Zdenek, et al, eds. Photographic atlas of the moon. New York, Academic Press, 1965.

Levitt, Israel M. & Marshall, Roy K. Star maps for beginners. New York, Simon & Schuster, 1964.

Moore, Patrick. The Atlas of the universe. Chicago, Rand McNally, 1970.

U.S. Geological Survey. Index map of subterrestial hemisphere of the moon. Washington, D.C., U.S.G.P.O., 1962--

CHEMISTRY

Pourbaix, Marcel, ed. Atlas of electrochemical equilibria in aqueous solutions. Elmsford, N.Y., Pergamon, 1966.

EARTH SCIENCES

Atkinson, R. C., et al. Atlas of landforms. New York, Wiley, 1966.

Barkley, Richard A. Oceanographic atlas of the Pacific Ocean. Honolulu, Univ. of Hawaii Pr., 1968.

Beutelspacher, H. & Van Der Marel, H. W. Atlas of electron microscopy of clay minerals and their admixtures. New York, American Elsevier, 1968.

Birch, T. W. Maps, topographical and statistical. 2nd ed. New York, Oxford Univ. Press, 1964.

Kerekes, C. Atlas & tables for emission spectrographic analysis of rare earth elements. Elmsford, N. Y., Pergamon, 1969.

Life Editors & Rand McNally. Atlas of the world. New York, Time-Life, 1968. (Life World Library Series.)

Maxwell, W. G. Atlas of the great barrier reef. New York, American Elsevier, 1969.

Rand McNally & Co. The International atlas. Text in English, German, Spanish and French. Chicago, Rand McNally, 1969.

Simpson, B. Geologic maps. Elmsford, N. Y., Pergamon, 1968. (Geology Division Series.)

Stevenson, Merritt R. Marine atlas of the Pacific coastal waters of South America. Berkeley, Cal., Univ. of Cal. Press, 1970.

Thoren, Ragnar. Picture atlas of the Arctic. New York, American Elsevier, 1969.

U. S. Environmental Data Service. Daily weather maps, weekly series. Washington, D. C., U. S. G. P. O., 1968-- (continues: U. S. Weather Bureau. Daily weather map. 1945--.)

U. S. Geological Survey. Index to geologic mapping in the United States. Washington, D. C., U. S. G. P. O., 1947-- (A series of state index maps.)

__________. Index to topographic maps of the United States. Washington, D. C., U. S. G. P. O., 1935-- (A series of state index maps.)

__________. Transcontinental geophysical maps series. Washington, D. C., U. S. G. P. O., 1968--

U. S. Library of Congress. A List of geographical atlases in the Library of Congress. Compiled by Clara E. LeGear. Washington, D.C., U. S. G. P. O., v. 5--, 1958--

U. S. National Aeronautics and Space Administration. Earth photographs from Gemini III, IV, and V. Washington, D. C., U. S. G. P. O., 1967. (NASA SP129.)

Upton, William Bayly, Jr. Landforms and topographic maps illustrating landforms of the continental United States. New York, Wiley, 1970.

Wahl, Eberhard W. and Lahey, James F. Seven hundred MB atlas for the northern hemisphere: five-day mean heights, standard deviations & changes. Madison, Wis., Univ. of Wis. Pr., 1969.

BIOLOGICAL SCIENCES

Barnes, Ervin H. Atlas & manual of plant pathology. New York, Appleton, 1968.

Bossy, Jean. Atlas of neuroanatomy and special sense organs. Philadelphia, Saunders, 1970.

Dodge, John D. Atlas of biological ultrastructure. New York, American Elsevier, 1968.

Eddy, Samuel. Atlas of drawings for chordate anatomy. 2nd ed. New York, Wiley, 1964.

__________, et al. Atlas of drawings for vertebrate anatomy. 3rd ed. New York, Wiley, 1964.

Etter, Lewis E. Atlas of roentgen anatomy of the skull. 3rd rev. ed. Springfield, Ill., C. C. Thomas, 1970.

Evans, F. Gaynor. Atlas of human anatomy. New York, Rowman & Littlefield, 1970.

Field, Hazel E. & Taylor, Mary E. Atlas of cat anatomy. Rev. ed. Ed. by Bernard D. Butterworth. Chicago, Univ. of Chicago Pr., 1969.

Ford, Donald H. & Schade, J. P. Atlas of the human brain. New York, American Elsevier, 1966.

Gilbert, Stephen G. Atlas of general zoology. Minneapolis, Burgess, 1965.

Hausman, Louis. Atlas of consecutive stages in the reconstruction of the nervous system. Springfield, Ill., C. C. Thomas, 1965.

Hsu, T. C. & Benirschke, K. Atlas of mammalian chromosomes. New York, Springer-Verlag, 1967-1969. 3 v.

Jordan, Emil Leopold. Animal atlas of the world. Maplewood, N. J., Hammond Inc., 1969.

Kemali, Milena & Braitenberg, Valentino. Atlas of the frog's brain. New York, Springer-Verlag, 1969.

Kiss, F. & Szentagothai, J. Atlas of human anatomy. Elmsford, N. Y., Pergamon, 1964. 3 v.

Laguens, Ruben P. & Gomez-Dumm, Cesar L. Atlas of human electron microscopy. St. Louis, Mosby, 1969.

Larsen, Hans-Walther. Manual and atlas of the fundus of the eye. Philadelphia, Saunders, 1969.

Miller, Richard A. & Burack, Ethel. Atlas of the central nervous system in man. Baltimore, Md., Williams & Wilkins, 1968.

Ogilvie, Alfred L. & Ingle, J. I. Atlas of pulpal & periapical biology. Philadelphia, Lea & Febiger, 1965.

Pernkopf, Edward. Atlas of topographical anatomy. Ed. by Ferner, Helmut. Philadelphia, Saunders, 1964. 2 v.

Piermattei, D. L. & Greeley, R. G. Atlas of surgical approaches to the bones of the dog & cat. Philadelphia, Saunders, 1966.

Roberts, Melville & Hanaway, Joseph. Atlas of the human brain in sections. Philadelphia, Lea & Febiger, 1970.

Scanga, Franco, ed. Atlas of electron microscopy; biological applications. New York, American Elsevier, 1964.

Sherwood, Nancy M. A Stereotaxic atlas of the developing rat brain. Berkeley, Cal., Univ. of Cal. Pr., 1970.

Sobotta, Johannes. Atlas of human anatomy. Ed. by Figge, Frank H. 8th ed. New York, Hafner, 1967-- 3 v. in 4 pts.

Spalteholz, Werner. Atlas of human anatomy. Ed. by Spanner, Rudolf. 16th ed. Tr. by Nederveen, Alexander. Philadelphia, F. A. Davis (Blackwell), 1967.

Szebenyi, Emil S. Atlas of Macaca mulatta. Rutherford, N. J., Fairleigh Dickinson Univ. Pr., 1970, c1969.

Waksman, Byron H. Atlas of experimental immunobiology and immunopathology. New Haven, Conn., Yale Univ. Press, 1970.

MEDICAL SCIENCES

Baronofsky, Ivan D. Atlas of precautionary measures in general surgery. St. Louis, Mosby, 1968.

Beattie, Edward J. & Economou, Steven. Atlas of advanced surgical techniques. Philadelphia, Saunders, 1968.

Becker, Walter, et al. Atlas of otorhinolaryngology & bronchoesophogology. Philadelphia, Saunders, 1969.

Bevelander, Gerrit. Atlas of oral histology & embryology. Philadelphia, Lea & Febiger, 1967.

Bjorkman, Nils. An Atlas of placental fine structure. Baltimore, Md., Williams & Wilkins, 1970.

Block, Matthew. Atlas of hematology. Philadelphia, Lea & Febiger, 1968.

Burch, Buford H. & Miller, Arthur C. Atlas of pulmonary resections. Springfield, Ill., C. C. Thomas, 1965.

Cozen, Lewis. Atlas of orthopedic surgery. Philadelphia, Lea & Febiger, 1966.

DeLand, Frank H. Atlas of nuclear medicine. v. 2, Lung and heart. Philadelphia, Saunders, 1970.

__________ & Wagner, Henry N., Jr. Atlas of nuclear medicine. v. 1, Brain. Philadelphia, Saunders, 1969.

Dison, Norma G. Atlas of nursing techniques. St. Louis, Mosby, 1967.

Forteza Bover, Geronimo & Candela, R. Baguena. Atlas of blood cytology, cytomorphology, cytochemistry, & cytogenetics. New York, Grune, 1964.

Freeman, William H. & Bracegirdle, Brian. Atlas of histology. New York, Dover, 1966.

Gass, John Donald M. Stereoscopic atlas of macular diseases. St. Louis, Mosby, 1970.

Gershon-Cohen, Jacob. Atlas of mammography. New York, Springer-Verlag, 1970.

Gross, Robert Edward. An Atlas of children's surgery. Philadelphia, Saunders, 1970.

Hayhoe, F. G. J. & Flemans, R. J. An Atlas of haematological cytology. New York, Wiley (Interscience), 1970.

Hershey, Falls B. & Calman, Carl H. Atlas of vascular surgery. 2nd ed. St. Louis, Mosby, 1967.

Iselin, Marc. Atlas of hand surgery. Tr. by Colwill. New York, McGraw-Hill, 1964.

Israels, M. C. Atlas of bone-marrow pathology. New York, Grune, 1967.

Janovski, Nikolas A. & Durbrousky, V. Atlas of gynecologic & obstetric diagnostic histopathology. New York, McGraw-Hill, 1968.

Jeffrey, Hugh C. & Leach, Robert M. Atlas of medical helminthology & protozoology. Baltimore, Md., Williams & Wilkins, 1966.

Kraus, Bertram S., et al. Atlas of developmental anatomy of the face. Scranton, Pa., Hoeber, 1966.

Larsen, Hans Walther. Manual and color atlas of the ocular fundus. Philadelphia, Saunders, 1969.

McDonald, George A., et al. Atlas of hematology. 2nd ed. Baltimore, Md., Williams & Wilkins, 1969.

Madden, John L. Atlas of technics in surgery. 2nd ed. New York, Appleton, 1964. 2 v.

Mark, D. D. & Zimmer, A. Atlas of clinical laboratory procedures, vol. 1, clinical chemistry. New York, McGraw-Hill, 1967.

Mori, Yoshitaka & Lennert, Karl. Electron microscopic atlas of lymph node cytology and pathology. New York, Springer-Verlag, 1969.

Moss, Emma Sadler & McQuown, Albert Louis. Atlas of medical mycology. 3rd ed. Baltimore, Williams & Wilkins, 1969.

Nicola, Toufick. Atlas of orthopaedic exposures. Baltimore, Md., Williams & Wilkins, 1966.

Parsons, Langdon & Ulfelder, Howard. Atlas of pelvic operations. 2nd ed., Philadelphia, Saunders, 1968.

Piliero, Sam J., et al. Atlas of histology. Philadelphia, Lippincott, 1965.

Reith, Edward & Ross, M. H. Atlas of descriptive histology. Scranton, Pa., Hoeber, 1970.

Semon, Henry C., et al. Atlas of the commoner skin diseases. Deer Park, N.Y., Brown Book Co., 1969.

Shirakabe, Hikoo. Atlas of X-Ray diagnosis of early gastric cancer. Philadelphia, Lippincott, 1967.

Thorek, Philip. Atlas of surgical techniques. Philadelphia, Lippincott, 1970.

Trethewie, E. R. Atlas of ABC electrocardiography. White Plains, N.Y., Phiebig, 1968.

Von Herrath, Ernest. Atlas of histology. Tr. by Kaysser, C. H. & Bartels, P. H. New York, Hafner, 1965.

White, R. R. Atlas of pediatric surgery. New York, McGraw-Hill, 1965.

Wilder, Joseph R. Atlas of general surgery. 2nd ed. St. Louis, Mosby, 1964.

Willson, J. Robert. Atlas of obstetric technic. 2nd ed. St. Louis, Mosby, 1969.

AGRICULTURAL SCIENCES

Annis, John R. & Allen, Algernon H. Atlas of canine surgery, vol. 1. Philadelphia, Lea & Febiger, 1967.

Getty, Robert. Atlas for applied veterinary anatomy. 2nd ed. Ames, Iowa, Iowa State Univ. Pr., 1964.

Yoshikawa, Tetuso. Atlas of brains of domestic animals. University Park, Pa., Pa. St. Univ. Pr., 1968.

Guides to the Literature

Factual questions such as who? what? when? where? or how much? require that a student locate biographical data, a formula, a definition, a fact, the date on which an event took place, an address or location, or verify a numerical value. The answers usually require only the reporting of information plus a bibliographical citation.

In any library there are usually many reference books which may prove helpful for answering factual questions. Nearly every book and magazine in a library could be considered a potential reference source. However, guides to the literature of a subject can aid students and research workers to make the best possible use of the standard reference books on a subject. They also serve as a reminder of the specialized dictionaries, cyclopedias, manuals, handbooks that have been published about a subject and as a cross reference to related topics. In a guide to the literature of a scientific subject specific book titles are listed which represent the kind of reference books needed to help answer factual questions.

One of the chief characteristics of science is that it is considered systematic; consequently there are obvious guides to the literature of the several different fields of the physical and biological sciences. Although guides to the literature can sometimes be out-of-date, they are useful for their general discussions and descriptions of standard reference books. Literature guides contain:

a) the most frequently used reference books and sources of information,

b) sources where additional information can be found,

c) sources of frequently used information, and

d) methods of using reference books to be informed about a subject.

SCIENCE, GENERAL

Asimov, Isaac. Intelligent man's guide to the physical sciences. New York, Washington Square Press, 1968.

Cheney, Frances Neel. Fundamental reference sources. Chicago, American Library Association, 1971.

Courtney, Winifred F., ed. The Reader's adviser: a layman's guide. 11th ed. New York, R. R. Bowker, 1969. 2 v.

Deason, Hilary J., comp. and ed. A Guide to science reading. New York, New American Library, 1966.

Grogan, Denis Joseph. Science and technology; an introduction to the literature. Hamden, Conn., Archon, 1970.

Guide to information sources in science and technology. New York, Wiley, v. 1-- 1963--

Guide to microforms in print. Washington, D. C., Microcard Editions, 1961--

Jenkins, Frances Briggs. Science reference sources. 5th ed. Cambridge, Mass., MIT Press, 1969.

Maichel, Karol. Guide to Russian reference books. Vol. V: Science, technology, and medicine. Stanford, Cal., Hoover Institution, Stanford Univ., 1967.

Malinowsky, Harold Robert. Science and engineering reference sources; a guide for students and librarians. Rochester, N. Y., Libraries Unlimited, 1967.

Scientific information notes: reporting national and international developments in scientific and technical information dissemination. Washington, D. C., National Science Foundation, v. 1-- Feb. /Mar. 1959--

Shaw, Ralph Robert. Pilot study on the use of scientific literature by scientists. Metuchen, N. J., Scarecrow Reprint Corp., 1956, reprint 1971.

Subject guide to microforms in print. Washington, D. C., Microcard Editions, 1962-1963--

UNESCO. UNESCO source book for science teaching. New York, UNESCO, 1969.

Winchell, Constance Mabel. Guide to reference books. 8th ed. Chicago, American Library Assn., 1967. Supplement, 1968--. compiled by E. P. Sheehy. (Appears in January and July issues of College & Research Libraries.)

MATHEMATICS

Morrill, Chester, Jr., ed. Computer and data processing information sources. Detroit, Gale Research Co., 1970.

Pemberton, John E. How to find out in mathematics: guide to sources of info. 2nd ed. rev. Elmsford, N.Y., Pergamon, 1969.

Pritchard, Alan. A Guide to computer literature. Hamden, Conn., Shoe String (Archon), 1969.

ASTRONOMY

Glasstone, Samuel. Sourcebook on the space sciences. New York, Van Nostrand-Reinhold, 1965.

Kemp, D. Alasdair. Astronomy and astrophysics; a bibliographical guide. Hamden, Conn., Archon Books, 1970.

PHYSICS

Glasstone, Samuel. Sourcebook on atomic energy. 3rd ed., New York, Van Nostrand-Reinhold, 1967.

Whitford, Robert H. Physics literature. 2nd ed. Metuchen, N.J., Scarecrow Press, 1968.

Yates, Bryan. How to find out about physics. Elmsford, N.Y., Pergamon, 1965.

CHEMISTRY

American Chemical Society. Literature of chemical technology. Washington, D.C., 1969. (Advances in chemistry

Series No. 78.)

Bottle, R. T., ed. The Use of chemical literature. 2nd ed. Hamden, Conn., Shoestring (Archon), 1969.

Burman, Charles R. How to find out in chemistry. 2nd ed. rev. Elmsford, N. Y., Pergamon, 1967.

Mellon, M. G. Chemical publications; their nature and use. 4th ed. New York, McGraw-Hill, 1965.

Signeur, Austin V. Guide to gas chromatography literature. New York, Plenum, 1964-1967. 2 v.

EARTH SCIENCES

Kaplan, S. R., ed. A Guide to information sources in mining, minerals, and geosciences. New York, Wiley, 1965.

U. S. Geological Survey. Guide to indexing bibliographies and abstract journals of the U. S. Geological Survey. Washington, D. C., 1967.

BIOLOGICAL SCIENCES

Asimov, Isaac. Intelligent man's guide to the biological sciences. New York, Washington Square Press, 1968.

Bottle, R. T. & Wyatt, H. V., eds. The Use of biological literature. Hamden, Conn., Shoe String (Archon), 1967.

Langman, Ida K., ed. A Selected guide to the literature on the flowering plants of Mexico. Philadelphia, Univ. of Pa. Pr., 1965.

Smith, Roger Cletus & Painter, Reginald H. Guide to the literature of the zoological sciences. 7th ed. Minneapolis, Burgess, 1967.

Swift, Lloyd H. Botanical bibliographies; a guide to bibliographic materials applicable to botany. Minneapolis, Burgess, 1970.

MEDICAL SCIENCES

Brunn, Alice Lefler. How to find out in pharmacy: a guide to sources of pharmaceutical information. Elmsford, N.Y., Pergamon, 1969.

Fox, David J. Fundamentals of research in nursing. 2nd ed. New York, Appleton, 1970.

Sanazaro, Paul J., ed. Current medical references. 6th ed. Los Altos, Cal., Lange Medical, 1970.

AGRICULTURAL SCIENCES

Parker, Dorothy & Carabelli, Angelina. Guide for an agricultural library survey for developing countries. Me- N.J., Scarecrow, 1970.

ENGINEERING SCIENCES

Anthony, L. J. Sources of information on atomic energy. Elmsford, N.Y., Pergamon, 1966.

Blaisdell, Ruth F., et al. Sources of information in transportation. Evanston, Ill., Northwestern Univ., 1964.

Bourton, Kathleen. Chemical and process engineering unit operations: a bibliographical guide. New York, Plenum, 1968.

Burkett, Jack & Plumb, P. How to find out in electrical engineering. Elmsford, N.Y., Pergamon, 1967.

Houghton, Bernard. Mechanical engineering: the sources of information. Hamden, Conn., Shoe String (Archon), 1970.

Kallard, Thomas. Holography; state-of-the-art review. Ansonia Station, N.Y., Optosonic Pr., 1970.

The Laser literature: an annotated guide. New York, Plenum, 1968. Covers literature 1963-1966, continuing Laser Abstracts (Plenum, 1964) which covered literature through mid-1963.

Passwater, Richard A. Guide to fluorescence literature. New York, Plenum (IFI), 1967-1970. 2 v.

Special Libraries Association. Guide to metallurgical information. Ed. by Tapia, E. W. & Gibson, E. B. 2nd ed. New York, 1966. (SLA Bibliography No. 3.)

Struglia, Erasmus J. Standards and specifications information sources; a guide to literature and to public and private agencies concerned with technological uniformities. Detroit, Gale Research Co., 1965.

Bibliographies

As the very next step after reading in the reference collection, you might consult a comprehensive published bibliography of your subject and see what additional materials you can discover in that way. But you could never afford to stop there, because a published bibliography may be a year or more out of date, so you will have to update it yourself by using your library's catalog as well as current periodical indexes. A published bibliography can sometimes be frustrating to use, because you will find listed in it so many titles which you would like to see, but cannot because your library does not own them. Even so, a published bibliography can give you some help if you are patient enough to trace out the titles listed in it which your library actually owns, or to search the holdings of other libraries.

Bibliographies, general and special, provide comprehensive listings and critical surveys of information sources. A bibliography is a list of references such as books, magazines or newspaper articles, manuscripts, documents, reports, etc. Often appearing as a reading list at the end of a chapter, at the end of a book as a list of sources consulted, or at the end of an encyclopedia article or reference book, bibliographies are one of the most valuable reference sources to be found in the library searcher's field of interest.

The use of prior bibliographies saves time and effort because they not only supply pertinent references but they usually supply the names of the leaders in the field and the most fruitful periodicals to be searched. They may list documents, pamphlets, brochures, theses and fringe publications that may not be covered by indexing or abstracting services. The elements usually contained in lists of descriptions of published materials include the author or authors; title; edition; place and date of publication, and name of publisher; and physical details (number of pages, presence of illustrations of one sort or another, and size). While these elements are usually regarded as the minimum essentials for the exact identification of published materials, bibliographies intended for special purposes, and those listing particular kinds of materials, will give more or less detail as required. However, the student compiling his own bibliography should regard the above elements as both minimal and sufficient for ordinary purposes. A special type of reference book needed to help understand bibliographies is a manual of bibliographical forms and citations. However, it is not suggested that a ready-made bibliography be substituted for an actual search and examination of the literature.

General bibliographies are helpful in a library search because they give a wide, though never complete, survey of the literature of many subjects and lead to information through the listing of specific books, or annotated descriptions of the book's contents. Bibliographies of bibliography have no limitations in showing the existence of books on many subjects by author.

National bibliographies list books published in a particular country or about a particular country.

Trade bibliographies furnish the record of printing output in a country and give descriptions of books in print plus cost and purchasing information not found in less comprehensive bibliographies. They should be consulted when their broad coverage of subjects can be most helpful.

However, there are reference books that are unique to each subject field, some whose titles are not obvious. Students who plan to specialize in a subject need to know that bibliographies of the subject are usually of two kinds: (a) basic bibliography, comprehensive up to a fixed date; (b) current bibliography, recording the literature of a given period, frequently one year at a time. In some cases a current bibliography connects exactly with the basic work and gives a comprehensive record for the whole field; for example, Nickles, Geologic Literature of North America, which covers publications to 1918, and is continued from 1919 by current bibliographies cumulating decennially. Together these furnish a full record of the literature of the subject.

Subject bibliographies are useful for (a) verification of incorrect or incomplete titles; (b) finding what material exists on a given topic; (c) an estimate of the value of a book or article, which may be given by an annotation or by reference to a critical review; (d) an abstract or digest of a particular book or article or note of its contents; (e) information on the fundamental or best books on a subject; (f) biographical data about an author.

Not all subject bibliographies will furnish information on all of these points; therefore, different types of bibliographies must be searched. For library searching the comprehensive subject bibliography which records books, periodical articles, and other analytical material, is the most useful. Within its stated limits it should give full and definite information about each item included, and should be so arranged and indexed that works can be found quickly by author, or by broad or specific subject.

If the bibliography is to serve as a critical guide to the literature, annotations or other indications of standing should be given. Frequently this type of information is best found in the selective bibliography which lists outstanding books and articles for a given subject. Other sources of subject bibliography are the reference lists on places, persons or things published by libraries, trade and professional societies. Closely related to bibliographies are indexes, abstracts, summaries, catalogs, and similar guides to published materials. The catalog is usually a listing of books which can be found together in a specific library. In almost all important areas of science and technology, new comprehensive bibliographies are being published.

SCIENCE, GENERAL

American book publishing record; BPR annual cumulative. New York, R. R. Bowker, 1960-- . Five year cumulative 1965-1969. New York, R. R. Bowker, 1970.

American reference books annual. Littleton, Colo., Libraries Unlimited, 1970. 2 v.

American scientific books. New York, R. R. Bowker, 1962-1965. (Continued in American Book Publishing Record. Annual. Cumulative. 1965--)

Announced reprints. Washington, D. C., Microcard Editions, 1969.

Ash, Lee & Lorenz, Denis. Subject collections: a guide to special book collections in libraries. 3rd ed. New York, R. R. Bowker, 1967.

Bertalan, Frank J. The Junior college library collection. Newark, N. J., Bro-Dart Pub. Co., 1970.

Besterman, Theodore. A World bibliography of bibliographies. 4th ed. New York, Rowman & Littlefield, 1970.

Bibliographic index; a cumulative bibliography of bibliographies. Bronx, N. Y., H. W. Wilson Co., 1938--. Annual and 3-yr. cumulations.

Book review digest. Bronx, N. Y., H. W. Wilson, v. 1, 1905--

Book review index. Detroit, Gale Research Co., v. 1, 1965--

Booklist; a guide to current books. (Formerly: Booklist and subscription books bulletin.) Chicago, American Library Assn., v. 1, 1905--.

Books in print: an author-title-series index to the Publishers trade list annual, New York, R. R. Bowker, 1948--

Choice (U. S.); books for the liberal arts curriculum by subjects. (Formerly: Choice: books for college libraries.) Chicago, American Library Assn., Assn. of College and Research Libraries Div., 1964--

Cumulative book index. New York, Wilson, 1898--

Deason, Hilary J. The A. A. A. S. science book list; a selected and annotated list of science and mathematics books for secondary school students, college undergraduates and nonspecialists. 3rd ed. Washington, D. C., A. A. A. S., 1970.

Directory of published proceedings. White Plains, N. Y., InterDok Corp., 1965--

Downs, R. B. & Jenkins, F. B., eds. Bibliography: current state and future trends. Urbana, Ill., Univ. of Ill. Pr., 1967. (Reprint of Library Trends, Jan. & Apr., 1967 with index.)

Forthcoming books. New York, R. R. Bowker, 1966.

McGraw-Hill Encyclopedia of Science & Technology Editors. Basic bibliography of science & technology. New York, McGraw-Hill, 1966. (Supplements Encyclopedia.)

Mapp, Edward, comp. Books for occupational education programs: a list for community colleges, technical institutes, and vocational schools. New York, R. R. Bowker, 1971.

Monthly checklist of state publications. Washington, D.C., U.S.G.P.O., 1910--

National Federation of Science Abstracting and Indexing Services. Proceedings. Washington, D.C., The Federation, 1970.

Publishers' weekly. New York, R. R. Bowker, 1872--

Reinhart, Bruce. The Vocational-technical library collection: a resource for practical education and occupational training. Williamsport, Pa., Bro-Dart Pub. Co., 1970.

Schutze, Gertrude. Documentation sourcebook. New York, Scarecrow Press, 1965. Supplement--Information and Library science sourcebook. Metuchen, N.J., Scarecrow, 1972.

Scholarly books in America. New York, American Univ. Press Services, Inc., 1959--

Sci-tech news. (Science-Technology Div., Special Libraries Assn.) Pasadena, Cal., Jet Propulsion Laboratory library, 1947--

Science books; a quarterly review. Washington, D.C., American Society for the Advancement of Science, v. 1, 1965--

Sonnenschein, William Swan. The Best books; a reader's guide and literary reference book, being a contribution towards systematic bibliography. 3rd ed. London, George Routledge and Sons, Ltd., 1910. Republished by Gale Research Co., Detroit, 1969. 6 v.

Strauss, Lucille J., et al. Scientific and technical libraries; their organization and administration. New York, Wiley (Interscience), 1964.

Subject guide to forthcoming books. New York, R. R. Bowker, 1967--

Technical book review index. New York, Special Libraries Assn., v. 1-- 1935--.

Titles in series; a handbook for librarians and students. Comp. by Eleanor A. Baer. 2nd ed. Metuchen, N.J., Scarecrow Pr., 1971. 2 v. 2nd supplement.

Turnbull, W. R. Scientific and technical dictionaries: an annotated bibliography. San Bernardino, Cal., Bibliotek, 1966-- v. 1, physical science and engineering, 1966. Supplement, 1968.

UNESCO. Bibliography of interlingual scientific & technical dictionaries. 5th ed. New York, Unipub, 1969. (Supplemented by listings in Bibliography, Documentation, Terminology.)

U.S. Government Publications. Monthly catalog. Washington, D.C., U.S.G.P.O., 1895--

Walsh, S. Padraig, ed. General encyclopedias in print. New York, R. R. Bowker, 1969.

MATHEMATICS

American Mathematical Society. New publications. Providence, R.I., American Mathematical Society. New Series, No. 1, 1964--

Computer literature bibliography. Ed. by Youden, W. W. Washington, D. C., U.S.G.P.O., 1965-- (v. 1: 1946-63; v. 2: 1964-67.) (SP No. 309.)

Gale Research Company. Statistical sources. 2nd ed. Detroit, Gale Research Co., 1966.

Lancaster, H. O. Bibliography of statistical bibliographies. New York, S-H Service Agency, 1968.

Schaaf, William Leonard. A Bibliography of recreational mathematics. 4th ed. Washington, D.C. National Council of Teachers of Mathematics, 1970.

Sommerville, Duncan M'laren Young. Bibliography of non-Euclidean geometry. 2nd ed. New York, Chelsea Pub., 1970.

Wold, Herman O. A., ed. Bibliography on time series and stochastic processes; an international project. Cambridge, Mass., M.I.T. Press, 1966.

ASTRONOMY

Bibliography of natural radio emission from astronomical sources. Ithaca, N.Y., Cornell Univ. Pr., v. 1, 1962--

PHYSICS

Atomindex. (Formerly: List of references on nuclear energy.) (Text in various languages.) Vienna, Austria, International Atomic Energy Agency, Div. of Scientific and Technical Information, v. 10, 1968. Available in microfiche.

Kuckowicz, B. Bibliography of the neutrino. New York, Gordon & Breach, 1967.

U.S. Atomic Energy Commission. Bibliographies of interest to the atomic program. Oak Ridge, Tenn., 1958-- (Title varies.)

U.S. Atomic Energy Commission. Technical books and monographs. Washington, D.C., 1st ed., 1959--

EARTH SCIENCES

American Society for Metals. Bibliography series. Metals Park, Ohio, A.S.M., v. 1, 1967--

Annotated bibliography of economic geology. (Geological Society of America; Society of Economic Geologists.) Lancaster, Pa., Economic Geology Pub. Co., v. 1-- 1929--

Annotated bibliography on hydrology and sedimentation. Washington, D.C., U.S.G.P.O., 1952-- (title varies; succeeds Bibliography of hydrology.)

Bibliography and index of geology. Boulder, Colo., Geological Society of America, v. 33, 1969-- monthly with an annual index. Successor to Bibliography of index of geology exclusive of North America. Boulder, Colo., Geological Society of America, v. 1-32, 1934-68.

Bibliography of fossil vertebrates. New York, Geological Society of America, 1902--
Literature to 1900 comp. by O. P. Hay (U.S.G.S. Bulletin 179), 1902.
1902-1927 comp. by O. P. Hay (Carnegie Institute Pub. No. 390, v. 1-2), 1929-30.
1928-1933 comp. by C. L. Camp, et al. (G.S.A. Spec. Paper 27), 1940.
1934-1938 comp. by C. L. Camp, et al. (G.S.A. Spec. Paper 42), 1942.
1939-1943 comp. by C. L. Camp, et al. (G.S.A. Memoir 34), 1949.
1944-1948 comp. by C. L. Camp, et al. (G.S.A. Memoir 57), 1953.
1949-1953 comp. by C. L. Camp, et al. (G.S.A. Memoir 84), 1961.
1954-1958 comp. by C. L. Camp, et al. (G.S.A. Memoir 92), 1964.
1959-1963 comp. by C. L. Camp, et al. (G.S.A. Memoir 117), 1968.

Oceanic abstracts. La Jolla, Cal., Oceanic Library and Information Center, v. 1-2: state-of-the-art--instrumentation (annotated bibliography), 1966-1967. See Oceanic Citation Journal.

Sparrow, Christopher J. & Healy, Terry R. Meteorology and climatology of New Zealand; a bibliography. New York, Oxford Univ. Pr. for Univ. of Auckland, N.Z., 1968.

U.S. Geological Survey. Bibliography of North American geology. Washington, D.C., U.S.G.S., 1896-- (Issued as U.S.G.S. Bulletins. Titles vary.)

__________. Publications for the geological survey, 1879-1961. Washington, D.C., U.S.G.P.O., Supplements (annual.) 1962-- (Supplemented monthly by "New Publications of the Geological Survey.")

U.S. Library of Congress. Bibliography on snow, ice, and permafrost with abstracts. Washington, D.C., v. 1, 1951-- (Title varies.)

_________. Soviet geography: a bibliography. Westport, Conn., Greenwood, 1968.

_________. United States IGY bibliography, 1953-1960; an annotated bibliography of United States contributions to the IGY and IGC (1957-1959). Washington, D.C., National Research Council, 1963. (N.R.C. Pub. No. 1087.)

Wheat, James Clement and Brun, Christian F. Maps and charts published in America before 1800; a bibliography. New Haven, Conn., Yale Univ. Pr., 1969.

BIOLOGICAL SCIENCES

Current bibliography for aquatic sciences and fisheries. Winchester, England, Taylor & Francis, 1964--

Huntia; a yearbook of botanical and horiticultural bibliography. Pittsburgh, Hunt Botanical Library. v. 1, 1964--

Lawrence, G. H. M., et al. Botanico-Periodicum-Huntianum. Pittsburgh, Hunt Botanical Library, 1968.

Wulff, L. Yvonne, et al. Physiological factors relating to terrestrial altitudes: a bibliography. Columbus, Ohio, Ohio State Univ. Library, 1968.

MEDICAL SCIENCES

Aerospace medicine and biology, a continuing bibliography. Washington, D.C., U.S. Library of Congress, v. 1, 1952-- (title varies.)

Alexander, Raphael, ed. Sources of medical information: a guide to organizations and government agencies which are sources of information in fields of medicine, health, disease, drugs, mental health and related areas, and to currently available pamphlets, reprints and selected scientific papers. New York, Exceptional Books, 1969.

Altman, Isidore, et al, eds. Methodology in evaluating the quality of medical care: an annotated selected bibliography. Pittsburgh, Pa., Univ. of Pittsburg Press, 1970. (Contemporary Community Health Series.)

American Academy of Orthopaedic Surgeons. Selective bibliography of orthopaedic surgery. 2nd ed. St. Louis, Mosby, 1970. Includes supplement.

Bibliography of the history of medicine. Washington, D.C., U.S. National Library of Medicine, v. 1, 1965--

Blake, John Ballard & Roos, Charles, eds. Medical reference works, 1679-1966; a selected bibliography. Chicago, Medical Library Assn., 1967. (Med. Lib. Assn. Pub. No. 3.) Supplement I, 1967. Supersedes the bibliographies published as part of the Medical Library Association's Handbook of Medical Library Practice, 2nd ed., 1956.

Current bibliography of epidemiology. New York, American Public Health Assn., 1969.

Dollen, Charles. Abortion in context: a select bibliography. Metuchen, N.J., Scarecrow, 1970.

The Drug scene, a bibliography. New Orleans, New Orleans Public Library, 1970.

Garrison, Fielding Hudson. A Medical bibliography; an annotated checklist of texts illustrating the history of medicine. Ed. by L. T. Morton. 3rd ed. Philadelphia, Lippincott, 1970.

Herskowitz, Irwin H. Bibliography on the genetics of drosophila; pt. 5. New York, Macmillan, 1969.

Menditto, Joseph. Drugs of addiction and non-addiction, their use and abuse; a comprehensive bibliography, 1960-1969. Troy, N.Y., Whitson Pub. Co., 1970.

Miller, Genevieve. Bibliography of the history of medicine in the United States and Canada, 1939-1960. Baltimore, Johns Hopkins Pr., 1964.

National Institutes of Health, Alerting Service, Virology Research Resources Branch, National Cancer Institute. Reading guide to the cancer-virology. Bethesda, Md., 1964--

National Institutes of Health, National Institute of Arthritis and Metabolic Diseases. Artificial kidney bibliography.

Washington, D. C., U. S. G. P. O., v. 1, 1967.

North Carolina University. School of Library Science. An Introduction to the literature of the medical sciences. 2nd ed., Chapel Hill, N. C., Univ. of N. C. Book Exchange, 1967.

Pasztor, Magda & Hopkins, Jenny. Bibliography of pharmaceutical reference literature. Philadelphia, Rittenhouse, 1968.

Physician's book compendium. New York, Physician's Book Compendium, 1969-1970.

Smoking and health bibliographical bulletin. (Formerly: Smoking and health bibliography.) Arlington, Va., Clearinghouse for Smoking and Health, v. 1, 1965--

Thornton, John Leonard. Medical books, libraries & collectors: a study of bibliography and the book trade in relation to the medical sciences. 2nd ed. London, British Book Centre, 1966.

Tissue culture bibliography. Bethesda, Md., Microbiological Assn., Inc., 1960--

U. S. National Library of Medicine. Monthly bibliography of medical reviews. Washington, D. C., U. S. G. P. O., v. 1, 1968-- Supersedes Bibliography of Medical Reviews, v. 1-12, 1955-1967. Also included in monthly issues of Index Medicus and cumulated in Cumulated Index Medicus.

__________. Toxicity bibliography. Washington, D. C., U. S. G. P. O., v. 1, 1968--

U. S. Public Health Service. Film reference guide for medicine and sciences. Washington, D. C., 1956-- (PHS Pub. 487 rev.)

U. S. Veterans Administration. Basic list of books and journals for Veterans Administration Medical Libraries. Washington, D. C., 1967. (G-14, M-2, Part XIII, rev.)

Wells, Dorothy P. Drug education: a bibliography of available inexpensive materials. Metuchen, N. J., Scarecrow, 1972.

AGRICULTURAL SCIENCES

Edwards, Everett Eugene. A Bibliography of the history of agriculture in the United States. New York, Burt Franklin, 1970. (Bibliography & Reference Series, No. 346.)

Schlebecker, John T. Bibliography of books & pamphlets on the history of agriculture in the United States, 1607-1967. Santa Barbara, Cal., ABC-Clio, 1969.

Smith, Charles L. Bibliography on the anti-monopoly 160-acre water law. Berkeley, Cal., Charles L. Smith, 1969.

U.S. Department of Agriculture. List of available publications of the United States Department of Agriculture. Washington, D.C., 1929-- (Updated by bimonthly list of publications and motion pictures, 1897--.)

__________. List of publications and patents with abstracts. Albany, Cal., Western Utilization Research & Development Div., 1955--

__________. Motion pictures of the U.S. Department of Agriculture. Washington, D.C., 1960-- (Updated by bimonthly list of publications and motion pictures, 1897--.)

U.S. National Agricultural Library. Div. of Indexing and Documentation. Bibliography of agriculture. Washington, D.C., v. 1, 1942--. (m. 11 times; Dec. a subject and author index.)

ENGINEERING SCIENCES

Allen, Ruth. An Annotated bibliography of biomedical computer applications. Detroit, Management Info. Services, 1970.

Ashburn, Edward V., ed. Laser literature; a permuted bibliography 1958-1966. North Hollywood, Cal., Western Periodicals Co., 1967. 2 v.

Codlin, E. M. Cryogenics and refrigeration; a bibliographical guide. New York, Plenum (IFI), 1968.

Cohn, M. Z. Limit design for reinforced concrete structures; an annotated bibliography. Detroit, American Concrete Institute, 1970. (ACI Bibliography No. 8.)

Current literature in traffic and transportation. Evanston, Ill., Northwestern University, v. 1, 1960--

Current papers in electrical and electronics engineering/CPE. New York, Institute of Electrical and Electronics Engineers, Inc., 1969.

Current papers on computer & control/CPC. New York, Institute of Electrical and Electronics Engineers, Inc., 1969.

Engineers' Council for Professional Development. Selected bibliography of engineering subjects. New York, 1937--. These bibliographies of books on engineering subjects include:

I. Mathematics, Mechanics & physics, 1962.
II. Aeronautical engineering, 1950.
III. Civil engineering, 1962.
IV. Ceramic engineering, 1965.
V. Metallurgical, mining & geological engineering, 1962.
VI. Mechanical engineering, 1955.
VII. Electrical engineering, 1958.
VIII. Chemical engineering, 1962.
IX. Industrial engineering, 1962.

Engineers Joint Council. Learning resources. New York, 1968.

Engineering Societies Library. Bibliography on filing, classification and indexing systems, and thesauri for engineering offices. New York, 1966.

Ferguson, Eugene S., ed. Bibliography on the history of technology. Cambridge, Mass., MIT Press, 1968. (Society for the History of Technology Series, No. 5.)

Mann, John Y., comp. Bibliography on the fatigue of materials, components, and structures, 1838-1950. New York, Pergamon for Royal Aeronautical Society, 1970.

Moore, C. K. & Spencer, K. J. Electronics: a bibliographical guide, vol. 2. New York, Plenum (IFI), 1966.

U. S. National Bureau of Standards. Publications of the National Bureau of Standards, 1901 to June 30, 1947. Washington, D. C., 1948. Supplementary lists, 1958--

Weiner, Jack & Roth, Lillian. Air pollution in the pulp and paper industry. Appleton, Wisc., Inst. of Paper Chem., 1969. (Inst. of Paper Chem. Biblio. Series No. 237.) Supplement #1, 1970.

Yescombe, E. R. Sources of information on the rubber, plastics and allied industries. Elmsford, N. Y., Pergamon, 1968. (International Series of Monographs on Library and Information Science, v. 7.)

Catalogs of Library Collections

The most convenient source of information on books about a subject is a library's own catalog. Because a library's subject collections are usually not complete or up-to-date, further search is necessary in other indexes to books and book collections.

It is essential in an academic library search to review the catalogs of the Library of Congress. They are unusually comprehensive and could be considered universal bibliographies since the Library of Congress is entitled by law to receive copies of all books copyrighted in the U.S. Its present collection numbers more than 30 million items, including over nine million pamphlets, over 11 million manuscripts, and millions of photographs, pieces of music, maps, and periodicals. For example, The National Union Catalog contains over 15 million cards representing over eight million titles. Complete cataloging information is included to enable the library searcher to select material for use and to copy book citations verbatim. By including monographs, new serial and periodical titles, technical reports, maps, documents and audiovisual materials, the Library of Congress catalogs, published annually, serve as reference books for students, educators and scientists. The Library of Congress also contains over a million book titles in Slavic, Hebraic, Japanese, and Chinese, making its catalogs a useful record of foreign language books. The major national card catalogs are available in book form. The National Union Catalog not only lists the major holdings of the Library of Congress but also those of over 700 cooperating American libraries by a system of letters prefixed by the state abbreviation; for example, LSU for Louisiana State University. Additional locations are listed in Symbols Used in the National Union Catalog of the Library of Congress (8th ed., Washington, D.C., The Library of Congress, 1960) which is frequently updated. Thus a current bibliographic key to a vast national collection is readily available to the library searcher.

There are many printed catalogs of academic and special research libraries that are useful for determining a book's location and the following bibliographical information: identification and verification of titles; information about authorship; description of books; edition and contents notes.

Book catalogs are usually assembled from the library's author, title, and subject catalog cards. Dictionary catalogs interfile entries by author, title, or subject in a single alphabetical arrangement. Classified catalogs, however, arrange the entries according to a classification scheme. Author, Title, and Report Number Indexes are sometimes included. Dictionary and subject catalogs of libraries or groups of libraries also serve as useful bibliographies because they give the location of at least one copy of each title listed. Regional union catalogs exist in Denver, Louisiana, Portland, Oregon, Philadelphia, Spokane and Seattle, Washington, and Texas.

Accession lists, similar to library book catalogs, may be a monthly or annual record of the publications acquired by a library covering a particular subject. Both commercial and government publications may be included. A few libraries print accession lists and distribute them on a subscription basis; others make mimeographed or similar duplicate copies for free distribution to a limited number of other libraries or individuals. These can be obtained by writing to the library.

GENERAL SCIENCE

John Crerar Library. Catalog. Boston, G. K. Hall, 1967. Author-title catalog. 35 v. Classified catalog. 42 v. Subject index to classified catalog. 1 v.

U. S. Library of Congress. National union catalog: a cumulative author list representing Library of Congress printed cards and titles by other American libraries. Washington, D. C., 1956--

__________. Subject catalog. Washington, D. C., 1950--

__________. Subject headings used in the dictionary catalogs. Supplements. Washington, D. C., 1908--

EARTH SCIENCES

United States Geological Survey Library, Department of the Interior. Catalog. Boston, G. K. Hall, 1965. 25 v.

University of California. Scripps Institution of Oceanography, La Jolla. Library. Catalogs of the Scripps Institution of Oceanography Library: author-title catalog. Boston, G. K. Hall, 1970. 7 v.

__________. Catalogs of the Scripps Institution of Oceanography Library: shelf list. Boston, G. K. Hall, 1970. 2 v.

__________. Catalogs of the Scripps Institution of Oceanography Library: shelf list of documents, reports and translations collection. Boston, G. K. Hall, 1970.

__________. Catalogs of the Scripps Institution of Oceanography Library: subject catalog. Boston, G. K. Hall, 1970. 2 v.

BIOLOGICAL SCIENCES

Catalogue of botanical books in the collection of Rachel McMasters Miller Hunt. Pittsburgh, Hunt Botanical Library, 1958--

Harvard University. Museum of Comparative Zoology. Library catalogue. Boston, G. K. Hall, 1968. 8 v.

Horticultural Society of New York. Printed books, 1481-1900, in the Horticultural Society of New York. Listing by Elizabeth Cornelia Hall. New York, The Society, 1970.

Massachusetts Horticultural Society. Library. Dictionary catalog. Boston, G. K. Hall, 1963. 3 v.

McGill University. Dictionary catalogue of the Blacker-Wood Library of Zoology & Ornithology. Boston, G. K. Hall, 1966. 9 v.

MEDICAL SCIENCES

Columbia University Medical Library. Recent acquisitions: a selected list and medical reference notes. Irvington-on-Hudson, N.Y., Series 3, No. 1, 1965--

New York Academy of Medicine. Author and subject catalogs of the Library of the New York Academy of Medicine. Boston, G. K. Hall, 1970. Author catalog, 43 v. Subject catalog, 34 v.

U. S. National Library of Medicine. Catalog. Washington, D. C., U. S. Library of Congress, 1948-1965. Annual with quinquennial cumulations, 1950-54--1966. (Title varies.) Continued with next listing.

_________. Catalog, current. Washington, D. C., U. S. G. P. O., v. 1, 1966--. Biweekly, cumulated quarterly; annually cumulated into bound volume.

_________. A Catalogue of Sixteenth Century printed books in the National Library of Medicine. Washington, D. C., U. S. G. P. O., 1967.

AGRICULTURAL SCIENCES

U. S. National Agricultural Library. Dictionary catalog of the National Agricultural Library, 1862-1965. Washington, D. C., v. 1, 1968. 73 v.

_________. Catalog. v. 1, 1966-- Monthly. Supplements dictionary catalog above.

Yale University. School of Forestry Library. Dictionary catalogue of the Henry S. Graves Memorial Library. Boston, G. K. Hall, 1962. 12 v.

ENGINEERING SCIENCES

Denver Public Library. New additions. Denver, Colo., Denver Public Library, Science and Engineering Dept., 1930--

Engineering Societies Library. Classed subject catalog. Boston, G. K. Hall, 1964. 13 v. Supplements, 1964--

New York Public Library. New technical books; a selected list on industrial arts and engineering added to the New York Public Library. New York, 1915--

Translations

When you have read all books in the English language in a subject, then you need to know about translations of books or articles published in foreign journals now available in English. To give some idea of the world wide scope of scientific and technical literature, it has been estimated that at least one million scientific articles, reports, patent specifications and books are added to the world's libraries every year, but over 50% of these are written in languages which more than 50% of the world's scientists cannot read. The importance of translating services becomes evident in the sciences when one considers that 90% of the medical literature is published in six languages besides English: German, French, Japanese, Italian, Spanish and Russian. In chemistry at least all of these languages plus Dutch are important. Furthermore, a good part of this material appears only in periodicals.

Translating requires many hours in patient research and the skillful copying of the necessary bibliographical data, including the transliteration of titles, and other information into Latin characters. The International Organization for Standardization (ISO) system for transliterating Cyrillic alphabets, used in Bulgarian, Byelorussian, Serbian, Ukranian and other languages, may be used to obtain uniformity in the spelling of authors' names in these languages. The transliteration of other languages using other alphabets is not yet based on uniform international systems.

With the development of extensive translation programs, domestic and foreign, government and private, it is possible to determine the existence, availability and location of translated scientific and technical reports, periodicals and books. Fortunately, information is centralized through the cooperation of the Clearinghouse for Federal Scientific and Technical Information (CFSTI) and a translation center maintained at the John Crerar Library, Chicago.

Much useful material is published in foreign language journals. The value of such journals is enhanced if they have English summaries or abstracts. Several indexes to articles and other materials that have already been translated are available. The major current index, Translations

Register Index, from the Special Libraries Association, is most helpful. English-language abstracting journals are also helpful in solving the language problem. Some foreign journals are published in side-by-side translations.

A number of government and private organizations now produce selective and cover-to-cover translations of journals, serials, books and reports published in Russian, Chinese, Japanese, and other languages. In larger research and special technical libraries a translation service may be available, either an individual or an organization to translate the needed information. The following bibliographies, guides and indexes to the regular translation of certain types and forms of documents and periodical are of help or available.

GENERAL SCIENCE

American translator. Ed. D. P. Moynihan. Text in English, Russian and various languages. New York, American Translators Assn., No. 2, 1968.

Gorokhoff, Boris Ivanovitch. Publishing in the USSR. Bloomington, Ind., Indiana University Press, 1959. (Russian and East European Series, v. 19.)

Himmelsbach, Carl J. & Boyd, Grace E., eds. A Guide to scientific and technical journals in translation. New York, Special Libraries Assn., 1968.

Horecky, Paul Louis. Libraries and bibliographic centers in the Soviet Union. Bloomington, Ind., Indiana University Press, 1959. (Russian and European series, v. 16.)

Index translationum. New York, UNESCO, v. 1-31, 1932-1940. New series: v. 1, 1948--

Kaiser, Francis E., ed. Translators and translations: services and sources. New York, Special Libraries Assn., 1965.

Maichel, Karol. Guide to Russian reference books, v. 5: Science, technology, medicine. Stanford, Cal., Hoover Institution on War, Revolution, and Peace, Stanford University, 1967. (Hoover Institution. Bibliographical series, v. 5.)

Neiswender, Rosemary, ed. Guide to Russian references and language aids, no. 4. New York, Special Libraries Assn., 1962.

Science East to West. Paris, n. 1, Feb. 1960-- (Title varies: Russian Technical Literature, n. 1-12, Feb. 1960-Dec. 1963.)

Technical translations. U.S. Department of Commerce, Springfield, Va., Clearinghouse, v. 1-- 1959--

Translations register index. New York, Special Libraries Assn., v. 1, 1967--

UNESCO. General catalogue of UNESCO publications and UNESCO sponsored publications. Paris, 1949-59, 1962. Supplements: 1960-63, 1964.

U.S. Library of Congress. Monthly index of Russian accessions. Washington, D.C., U.S.G.P.O., v. 1-- 1948-- "a record of the publications in the Russian language issued in and outside the Soviet Union that are currently received in the Library of Congress and a group of cooperating libraries." (Text in English and Russian.)

U.S. Library of Congress. Science and Technology Division. List of Russian serials being translated into English and other Western languages. 1st ed. Washington, D.C., 1960-- (Annual.) Covering: Foreign periodical bibliography. Out of print. Order from U.S. Library of Congress, Photoduplication Service (Entry 37990). Languages: French, German, Russian. Subjects: agr. sci, astron, biol, biochem, biophys, entomol, med, pharmacol, physiol, psychol, bot, chem, crystallog, inorg. chem, org. chem, phys. chem, geod, geog, geol, oceanog, eng, cer. eng, fuels, plast, rub, text, elec. eng, electron, mech. eng, met. eng, math. sci, instrumtn, met, phys, acoust, mechs, opt, aeronauts, at. ener, automatic control, cybernetics, foundry practice, geochem, glass technol, pats, nuclear phys, pathol, petro, radio sci, snow mechs, soils, spectroscopy, telecommunics, welding, wood ind.

World index of scientific translations. Delft, Netherlands, European Translations Center, v. 1-- 1967--

MATHEMATICS

American Mathematical Society. Translations. v. 1, 1955--

PHYSICS

Acta physica sinica abstracts. New York, American Institute of Physics, v. 1-- 1965--

Physics express. New York, International Physical Index, Inc., v. 1-- 1958--

Selected abstracts of non-U.S. literature on production and industrial uses of radioisotopes. (Formerly: Selected abstracts of foreign literature). Oak Ridge, Tenn., Oak Ridge National Laboratory, v. 1-- 1966--

Transatom bulletin; a monthly guide to nuclear literature in translation (text in English). Ed. EURATOM-CID. Brussels, Belgium, Agence et Messageries de la Presse, 1960--

CHEMISTRY

Reid, Ebenezer Emmet. Chemistry through the language barrier; how to scan chemical articles in foreign languages with emphasis on Russian and Japanese. Baltimore, Johns Hopkins, 1970.

ENGINEERING SCIENCES

Automation express. New York, International Physical Index, Inc., v. 1-- 1958--

Electronics express. New York, International Physical Index, Inc., v. 1-- 1958--. Comprehensive digest of current Russian literature dealing with electronic topics.

Review Journals

It is essential for anyone who embarks on the study of a subject for purely practical or for scientific purposes to become thoroughly familiar with the practical know-how and the scientific results which have been accumulated in the subject. Current events, new developments and trends in most fields of science and technology are summarized in a group of reference books known as reviews. They are published annually to provide authoritative critical reviews and comprehensive summaries of recent research in scientific and technological subjects.

Unlike the abstract of a periodical article which usually furnishes facts contained in the article rather than an evaluation of those facts, the review covers all the significant articles reporting progress or research in a particular subject. The annual review is more comprehensive than the periodical article, but less exhaustive than the book which treats every aspect of a large subject. Information is presented in a form that enables the library searcher to determine quickly whether he should read further in the periodical articles listed in the bibliographies after each review article.

After World War II, most of the papers and articles dealing with scientific theory as well as the practical aspect of technological processes were widely scattered throughout various journals and manuals. Little had been done toward integrating the findings. Textbooks usually covered one topic, while handbooks became dated by rapid progress.

For these reasons annual reviews were developed to summarize established knowledge and critically disclose gaps of scientific knowledge. Nearly every major academic subject now has such an information service.

GENERAL SCIENCE

Annual review of information science and technology. Chicago, Encyclopaedia Britannica, v. 1, 1966--

MATHEMATICS

Advances in computers. New York, Academic Press, v. 1, 1960--

Advances in mathematics. New York, Academic Press, v. 1, 1961--

Annals of mathematics. Princeton, N.J., Princeton Univ. Pr., v. 1, 1884--

Progress in mathematics. New York, Plenum, v. 1, 1968--

ASTRONOMY

Advances in astronomy and astrophysics. New York, Academic Press, v. 1, 1962--

Advances in the astronautical sciences. New York, American Astronautical Society, v. 1, 1957--

Annual review of astronomy and astrophysics. Palo Alto, Cal., Annual Reviews, Inc., v. 1, 1963--

PHYSICS

Advances in atomic and molecular physics. New York, Academic Press, v. 1, 1965--

Advances in chemical physics. New York, Wiley (Interscience), v. 1, 1958--

Advances in chromatography. New York, M. Dekker, v. 1, 1965--

Advances in electronics and electron physics. New York, Academic Press, v. 1, 1948-- Supplements, 1963--

Advances in magnetic resonance. New York, Academic Press, v. 1, 1965--

Advances in nuclear physics. New York, Plenum Press, v. 1, 1968--

Advances in physics. London, Taylor and Francis Ltd., v. 1, 1952--

Advances in plasma physics. New York, Wiley (Interscience), v. 1, 1968--

Advances in theoretical physics. New York, Academic Press, v. 1, 1965--

Annual review of nuclear science. Palo Alto, Cal., Annual Reviews Inc., v. 1, 1952--

Applied spectroscopy reviews. New York, M. Dekker, v. 1, 1967--

Progress in nuclear energy. Elmsford, N.Y., Pergamon, v. 1, 1956-- (published in 12 series at irregular intervals.)

Reports on progress in physics. New York, American Institute of Physics, v. 1, 1934--

Reviews of plasma physics. New York, Plenum, v. 1, 1965--

Solid state physics: advances in research and application. New York, Academic Press, v. 1, 1955--. Supplements, 1958--

CHEMISTRY

Advances in analytical chemistry and instrumentation. New York, Wiley (Interscience), v. 1, 1960--

Advances in chemistry series; a continuing series of books published by the American Chemical Society. Washington, D.C., v. 1, 1950-- (Early titles available in microfiche only.)

Advances in heterocyclic chemistry. New York, Academic Press, v. 1, 1963--

Advances in high temperature chemistry. New York, Academic Press, v. 1, 1967-

Advances in inorganic chemistry and radiochemistry. New

York, Academic Press, v. 1, 1959--

Advances in macromolecular chemistry. New York, Academic Press, v. 1, 1968--

Advances in organic chemistry; methods and results. New York, Wiley (Interscience), v. 1, 1960--

Advances in organometallic chemistry. New York, Academic Press, v. 1, 1964---

Advances in photochemistry. New York, Wiley, v. 1, 1963--

Advances in physical organic chemistry. New York, Academic Press, v. 1, 1963--

Advances in polymer science. New York, Springer-Verlag, v. 1, 1958--

Advances in quantum chemistry. New York, Academic Press, v. 1, 1964--

Annual reports in medicinal chemistry. Sponsored by the Div. of Medicinal Chemistry, American Chemical Society. New York, Academic Press, v. 1, 1965--

Annual review of biochemistry. Palo Alto, Cal., Annual Reviews, Inc., v. 1, 1932--

Annual review of physical chemistry. Palo Alto, Cal., Annual Reviews, Inc., v. 1, 1950--

Chemistry and physics of carbon; a series of advances. New York, M. Dekker, v. 1, 1965--

Electroanalytical chemistry; a series of advances. New York. M. Dekker, v. 1, 1966--

Perspectives in structural chemistry. New York, Wiley, v. 1, 1967--

Progress in inorganic chemistry. New York, Wiley (Interscience), v. 1, 1959--

Progress in physical organic chemistry. New York, Wiley (Interscience), v. 1, 1963--

Progress in solid state chemistry. Elmsford, N.Y., Pergamon, v. 1, 1964--

Reviews in macromolecular chemistry. New York, Academic Press, v. 1, 1966--

Survey of progress in chemistry. New York, Academic Press, v. 1, 1963--

Techniques and methods of organic and organometallic chemistry. New York, M. Dekker, v. 1, 1969--

Transition metal chemistry; a series of advances. New York, M. Dekker, v. 1, 1965--

EARTH SCIENCES

Advances in environmental sciences. New York, Wiley (Interscience), v. 1, 1969--

Advances in geology. New York, Academic Press, v. 1, 1965--

Advances in geophysics. New York, Academic Press, v. 1, 1952--

Advances in hydroscience. New York, Academic Press, v. 1, 1964--

Annals of the International Geophysical Year. Elmsford, N.Y., Pergamon, v. 1, 1957--

Annals of the IQSY. Cambridge, Mass., MIT Press, v. 1, 1968--

Progress in oceanography. Elmsford, N.Y., Pergamon, v. 1, 1963--

BIOLOGICAL SCIENCES

Advances in applied microbiology. New York, Academic Press, v. 1, 1959--

Advances in biological and medical physics. New York, Academic Press, v. 1, 1948--

Advances in cell biology. New York, Appleton, v. 1, 1970--

Advances in ecological research. New York, Academic Press, v. 1, 1962--

Advances in genetics. New York, Academic Press, v. 1, 1947--

Advances in human genetics. New York, Plenum, v. 1, 1970--

Advances in morphogenesis. New York, Academic Press, v. 1, 1961--

Advances in oral biology. New York, Academic Press, v. 1, 1964--

Advances in virus research. New York, Academic Press, v. 1, 1953--

Annual review of ecology. Palo Alto, Cal., Annual Reviews, Inc., v. 1, 1970--

Annual review of entomology. Palo Alto, Cal., Annual Reviews, Inc., v. 1, 1956--

Annual review of genetics. Palo Alto, Cal., Annual Reviews, Inc., v. 1, 1967--

Annual review of microbiology. Palo Alto, Cal., Annual Reviews, Inc., v. 1, 1947--

Annual review of physiology. Palo Alto, Cal., Annual Reviews, Inc., v. 1, 1950--

Annual review of phytopathology. Palo Alto, Cal., Annual Reviews, Inc., v. 1, 1963--

Annual review of plant physiology. New York, Academic Press, v. 1, 1950--

Biological reviews of the Cambridge Philosophical Society. New York, Cambridge Univ. Press, v. 1, 1923--

Contributions to sensory physiology. New York, Academic Press, v. 1, 1965--

Current topics in developmental biology. New York, Academic Press, v. 1, 1966--

Evolutionary biology. New York, Appleton, v. 1, 1967--

International review of cytology. New York, Academic Press, v. 1, 1952--

International review of neurobiology. New York, Academic Press, v. 1, 1959--

Life sciences and space research. New York, Humanities, v. 1, 1962--

Neurosciences research. New York, Academic Press, v. 1, 1968--

Progress in biophysics and molecular biology. Elmsford, N.Y., Pergamon, v. 1, 1950-- (Former title of this annual: Progress in biophysics and biophysical chemistry.)

Progress in nucleic acid research and molecular biology. New York, Academic Press, v. 1, 1963--

Survey of biological progress. New York, Academic Press, v. 1, 1949--

MEDICAL SCIENCES

Advances in clinical chemistry. New York, Academic Press, v. 1, 1958--

Advances in immunology. New York, Academic Press, v. 1, 1961--

Advances in internal medicine. Chicago, Year Book Medical Pub., v. 1, 1942--

Advances in metabolic disorders. New York, Academic Press, v. 1, 1964--

Advances in pharmacology and chemotherapy. New York, Academic Press, v. 1, 1962--

Advances in surgery. Chicago, Year Book Medical Pub.,

v. 1, 1965--

Annual review of medicine. Palo Alto, Cal., Annual Reviews, Inc., v. 1, 1950--

Annual review of pharmacology. Palo Alto, Cal., Annual Reviews, Inc., v. 1, 1961--

International review of experimental pathology. New York, Academic Press, v. 1, 1962--

Medical opinion and review. New York, Medical Opinion and Review, v. 1, 1965--

Methods in cancer research. New York, Academic Press, v. 1, 1967--

Methods in medical research. Chicago, Year Book Medical Pub., v. 1, 1948--

Pathology annual. New York, Appleton, v. 1, 1966--

Pharmacological reviews. American Society for Pharmacology and Experimental Therapeutics. Baltimore, Md., Williams & Wilkins, v. 1, 1949--

Physiological reviews. Baltimore, American Physiological Society, v. 1, 1921--

Progress in clinical cancer. New York, Grune & Stratton, v. 1, 1965--

Progress in clinical pathology; a review of significant advances in the field of clinical pathology. New York, Grune & Stratton, v. 1, 1966--

Progress in experimental tumor research. White Plains, N.Y., Phiebig, v. 1, 1960--

Progress in medical virology. White Plains, N.Y., Phiebig, v. 1, 1958--

Progress in neurological surgery. Chicago, Year Book Medical Pub., v. 1, 1966--

Progress in surgery. White Plains, N.Y., Phiebig, v. 1, 1961--

Review of medical microbiology. Los Altos, Cal., Lange Med. Pubns., v. 1, 1954--

AGRICULTURAL SCIENCES

Advances in food research. New York, Academic Press, v. 1, 1948--

Advances in veterinary sciences. New York, Academic Press, v. 1, 1953--

Review of textile progress. New York, Plenum, v. 1, 1949--

ENGINEERING SCIENCES

Advances in applied mechanics. New York, Academic Press, v. 1, 1948-- Supplements, 1961--

Advances in biomedical engineering and medical physics. New York, Wiley, v. 1, 1968--

Advances in chemical engineering. New York, Academic Press, v. 1, 1948--

Advances in cryogenic engineering. New York, Plenum, v. 1, 1960--

Advances in electrochemistry and electrochemical engineering. New York, Wiley (Interscience), v. 1, 1961--

Advances in heat transfer. New York, Academic Press, v. 1, 1964--

Advances in information systems science. New York, Plenum, v. 1, 1969--

Advances in microwaves. New York, Academic Press, v. 1, 1966-- Supplement 1, 1970.

Advances in space science and technology. Ed. Frederick I. Ordway II. New York, Academic Press, 1959-1968. 9 v. Supplements.

Supplement #1: Space carrier vehicles: design, development & testing of launching rockets. Ed. by Lange,

O. H. & Stein, R. J., New York, Academic Press, 1963.

Supplement #2: Lunar & planetary surface conditions. Ed. by Weil, N. A. New York, Academic Press, 1965.

Annual safety education review. Washington, D. C., American Association for Health, Physical Education & Research, v. 1, 1962--

Applied mechanics review. Easton, Pa., American Society of Mechanical Engineers, v. 1, 1948--

Modern chemical engineering. New York, Van Nostrand-Reinhold, v. 1, 1963-- (a monographic series.)

Progress in astronautics and aeronautics. New York, Academic Press, v. 1, 1960--

Progress in separation & purification. New York, Wiley (Interscience), v. 1, 1968--

Simonyi, K. Foundations of electrical engineering. New York, Macmillan, 1963--

U. S. aircraft, missiles and spacecraft. Washington, D. C., National Aerospace Education Council, 1957-- (Annual.)

U. S. Bureau of Customs. Merchant vessels of the United States. Washington, D. C., U. S. G. P. O., v. 1, 1866-- (Annual.)

World review of nutrition and dietetics. White Plains, N. Y., Phiebig, v. 1, 1959--

Lists of Periodicals

In searching the resources of a library, periodicals, journals and serials can supplement and update the information found in reference books. A periodical is a serial publication issued in parts which usually contains articles by several named contributors; it generally has only one distinctive title and the successive parts or numbers are intended to appear at stated or regular intervals indefinitely. A periodical usually is published two or more times each year. There are general periodicals that publish only brief reports of original research; others, such as academy or society transactions, specialize in exhaustive scholarly and scientific reviews of work that is in an active state of development. Many scientific journals are issued as proceedings or transactions of professional scientific societies or are sponsored by such groups; some issue from government bureaus; others are published as independent enterprises. The important science journals have distinctive characters and maintain definite editorial policies over long periods of time. Serials are publications issued in successive parts, usually at regular intervals, and as a rule, intended to be continued indefinitely. Serials include periodicals, annuals (reports, yearbooks, etc.) and memoirs, proceedings, and transactions of societies.

Ideally, a science collection should have available all the periodicals in which scientific papers and articles are written, plus all back volumes of such journals since 1900 or earlier. The growth of science literature can be shown by comparing the 25,000 entries in the 1925-27 edition of the World List of Scientific Periodicals with the 36,000 entries in the 1934 edition and over 50,000 entries in the 1952 edition. The number has grown steadily until there are over 100,000 scientific and technical journals currently published. It has been estimated that it would take over ten years to read all the information published each and every year. Periodicals present a continuous record of scientific advance, though not as a story that is easily read since related information is frequently scattered in numerous articles. The larger the number of periodicals available, particularly of long runs of leading periodicals, the more information will be accessible. It is for this reason that periodicals and journals, in contrast to most books, appreciate in value with time.

The great advantage of science periodicals over textbooks is their frequency and regularity of issue, for it usually takes less time for authors to write up-to-date articles than a book. Consequently, periodicals contain the most recent available information regarding the immediate results of experimental research and announcements of technical developments. Scientific articles usually cover one subject in depth, describing methods, apparatus, and results. Usually, it takes a few years or more before the information contained in periodical articles is rewritten into textbooks, reviews and cyclopedias. The footnotes and bibliographies listed in the most recent articles serve as a primary source of information on a specific science subject. Periodicals also contain current subject bibliographies, the latest market prices, advertisements, and news items which are seldom covered in abstract and indexing publications.

Periodicals which are indexed annually, or whose articles are covered by an indexing service, are helpful in retrospective searching. Annual periodical indexes may appear in any one of several places: 1) fastened in the last issue of a periodical volume, or 2) with an issue of a succeeding periodical volume, or 3) bound separately. For some periodicals no index is provided, but a table of contents for the complete volume is bound at the front of the volume instead.

There are several lists or directories of periodicals which give information about periodicals, their correct titles, price and publisher. Consultation of one or more of the following lists will help identify periodicals for the various scientific and technological subjects.

PERIODICALS LISTS

Access. Washington, D.C., American Chemical Society, v. 1, 1969-- Replaces Chemical Abstracts, Lists of periodicals with key to library files, v. 1, 1962 and supplements 1962--.

American Chemical Society. Bibliography of chemical reviews. Washington, D.C., v. 1, 1960--

American National Standards Institute. American National Standard for the abbreviation of titles of periodicals. New York, 1969.

American Society for Testing and Materials. Coden for 22,544 titles of serials. 2nd supplement. Philadelphia, 1969.

Bonn, George S., ed. Japanese journals in science and technology: an annotated check list. New York, New York Public Library, 1960.

Brown, Peter & Stratton, George B. World list of scientific periodicals published in the years 1900-1960. Hamden, Conn., Shoe String (Butterworth), 1964. 3 v.

Geophysical abstracts, list of periodicals. Washington, D.C., United States Geological Survey, v. 1, 1961--

International Association of Agricultural Librarians and Documentalists. Current agricultural serials, a world list of serials in agriculture and related subjects (excluding forestry and fisheries.) Oxford, Alden. Current in 1964. 1965-67. 2 v.

List of periodicals indexed in Index Medicus. Washington, D.C., U.S. National Library of Medicine, v. 1, 1963--

Martyn, John and Gilchrist, Alan. An Evaluation of British scientific journals. London, ASLIB, 1969. (ASLIB Occasional Pub. No. 1.)

Massachusetts Institute of Technology Library. Current serials and journals in the MIT Libraries. Cambridge, Mass., 1st ed., 1957--

Meteorological and geoastrophysical titles. Boston, American Meteorological Society, v. 1, 1961--

N. W. Ayer & Son's directory of newspapers and periodicals. Philadelphia, Pa., N. W. Ayer & Sons, Inc., 1869--

National Science Foundation. A Guide to East European scientific and technical literature. Washington, D.C., 1963.

__________. List of Russian scientific journals available in English. Washington, D.C., N.S.F., Office of Scientific Information, August 1961, revised as necessary.

Serials bulletin. Ann Arbor, Mich., Univ. Microfilm

Service, 1969.

Tompkins, Mary L., comp. & ed. MAST: Minimum abbreviation of serial titles: mathematics. North Hollywood, Cal., Western Periodicals, 1969.

Tunevall, Gosta, ed. Periodicals relevant to microbiology and immunology--a world list--1968. New York, Wiley (Interscience), 1969.

Ulrich's international periodicals directory; a classified guide to current periodicals, foreign and domestic. 14th ed., New York, R. R. Bowker, 1971/72. 2 v. Supplement to 13th ed., 1970.

U.S. Library of Congress. Serial publications of the Soviet Union, 1939-1957: a bibliographic checklist. Washington, D.C., U.S. Library of Congress, Cyrillic Bibliographic Project, 1958.

U.S. Library of Congress, Science and Technology Division. A List of scientific and technical serials currently received by the Library of Congress. Washington, D.C., U.S.G.P.O., 1960.

U.S. National Agricultural Library. Serial publications indexed in Bibliography of agriculture. Rev. ed., Washington, D.C., 1965.

Vital notes on medical periodicals. Chicago, Medical Library Assn., v. 1, 1952--

Wall, Edward C. Periodical title abbreviations. Detroit, Gale Research Co., 1969.

Union Lists of Periodicals

A union list of periodicals is similar to a union book catalog arranged alphabetically by title of serials, periodicals, newspapers, or microforms to be found in the libraries of a city, region or country. The list supplies periodical title, place of publication and where current issues and back sets of the periodicals included in the list may be found. Union lists may be general or limited to a subject; details are usually given concerning title, changes of title, place of publication, date of founding, and of last volume if publication has ceased, and volume numbers. These lists can be of great help in locating periodicals in other libraries.

For many years libraries have depended upon the various union lists of serials as principal sources of information regarding the serial resources of libraries in the United States. Although such lists--notably the Union List of Serials of the H. W. Wilson Co., 1943--have rendered indispensable service of this kind, their manner of compilation has not enabled them to meet the problem of providing comprehensive information about newly available serials. In an effort to answer this need, the Library of Congress began publication in 1951 of Serial Titles Newly Received. But since this listing represented the receipts of only one library, it could not provide the desired degree of comprehensiveness or locate materials in other libraries.

Consequently, in order to afford fuller coverage of serials new to American libraries, the Joint Committee on the Union List of Serials recommended that Serial Titles Newly Received be expanded to serve as a continuing supplement to the Union List of Serials in the matter of new titles. Serial Titles Newly Received was discontinued, therefore, with its 1952 annual cumulation and was superseded in January 1953 by New Serial Titles which lists serials first published after December 31, 1949, and received by the Library of Congress and cooperating libraries. The titles previously listed in Serial Titles Newly Received which fall within the scope of New Serial Titles are also listed in the latter.

New Serial Titles appears in twelve monthly issues and in annual cumulations which have been in turn cumulated over five- or ten-year periods. For the period 1966-1969 the complete set of New Serial Titles will consist of the 1950-1960 and the 1961-1965 cumulative volumes as supplemented by the latest annual cumulation and monthly issues. New reports of library locations are published in each cumulative volume. However, in cases where the published locations of a given serial prove to be inadequate, the Serial Record Division of the Library of Congress will supply information concerning any additional locations reported to it. The entries in the monthly issues of New Serial Titles are also issued in subject sequence in the related publication, New Serial Titles--Classed Subject Arrangement.

In general, New Serial Titles holdings are shown in the summarized form used in the Union List of Serials. The National Union Catlog symbols are used to designate reporting libraries. A dash (--) at the end of any holding indicates that the library expects to continue to receive issues of the title. An asterisk (*) indicates that the library acquires a monograph series on a selective rather than a continuing basis. "CURRENT ISSUES" indicates that the library retains only a current file of the publication; "NOT RETAINED" that it either discards the publication or retains only a sample issue. Newspapers, looseleaf publications, books in parts, municipal government serial documents, publishers' series, motion pictures, filmstrips, and phonorecords are excluded. No holdings are shown for United Nations and United States Federal and State serial documents. For a list of U.N. depository libraries see the latest revision of Consolidated Lists of Depository Libraries and Sales Agents and Offices of the United Nations and the Specialized Agencies, issued by the U.N. Library in the U.N. Documents Series. For a list of Federal depository libraries see the most recent September issue of the Monthly Catalog of United States Government Publications, issued by the Superintendent of Documents. State documents are held by the Library of Congress and, generally, by the State Library or the issuing agency and the Midwest Inter-Library Center.

However, wanted books or periodicals can only be borrowed on interlibrary loan as a courtesy from the lending library. The lending library may not choose to honor an interlibrary loan request if the particular book is needed by its own patrons. Even when periodicals are not available for interlibrary loan, the knowledge of the location of particular volumes or sets is useful because of the possibility of obtaining photocopies of needed articles.

UNION LISTS OF PERIODICALS

British union-catalogue of periodicals, new periodical titles. London, Butterworths, 1964--

New Serial titles; a union list of serials commencing publication after December 31, 1949. (Alphabetical arrangement.) Includes annual cumulation. Washington, D.C., U.S. Library of Congress & Card Division, Navy Yard Annex. (1961-1965 cumulation published by New York, R. R. Bowker Co.)

New serial titles, classed subject arrangement. Washington, D.C., U.S. Library of Congress, Card Division, Navy Yard Annex, 1955--

New York State union list of serials. Prepared under direction of the New York State Library. New York, CCM Information Corp., 1970. 2 v.

Union list of scientific serials in Canadian libraries. 3rd ed. Ottawa, Canada, National Science Library, National Research Council of Canada, 1969.

Union list of serials -- education, science, medicine -- in the libraries of the University of Rochester. Rochester, N.Y., Univ. of Rochester Library, Information Systems Office, 1969. 2 v.

Other lists of periodicals will be found in the various periodical indexes. For example, Readers' Guide to Periodical Literature would help locate additional sources of information in general periodicals. It is helpful to remember that an indexed periodical has more reference value in a library search than one that is not indexed. A most convenient list of current periodicals is Applied Science and Technology Index. Titles are listed in alphabetical order of their abbreviated form used in the index and give the full title of the periodical and the name and address of the publisher.

Chapter III

HOW TO USE A LIBRARY'S MAIN CARD CATALOG

The Main Card Catalog is a record of all the books in a library. Learn how to use it, for it is the key to all the circulating and reference books, periodicals and library materials available for your use. Just as a telephone directory has two parts: (1) the list of names, and (2) the classified section, so the library's divided card catalog has two sections of cards: (1) Author-Title and (2) Subject Headings. In many libraries, however, these sections are combined into one alphabetical sequence.

Author-Title Catalog

Use this catalog to find the call number of a particular book or magazine, if you know its author or title. Looking for a book under its author's name is considered a good approach, as the author card is the main entry; that is, it may contain more extensive information about the book than do title or subject heading cards. The one exception to the separation of all Subject Heading entries from the Author-Title entries is: all subject headings of personal surnames are filed in the Author-Title section of the divided card catalog.

Cards for all books by one author are arranged alphabetically by title under the author's last name. Authors with the same last name are arranged alphabetically by their first names. Authors are usually individuals, but may be companies, institutions, governments, agencies, or other organizations. Examples: American Chemical Society; Massachusetts Institute of Technology; Society for Experimental Biology (Gt. Brt.); U.S. Department of Commerce; U.S. Dept. of Health, Education, and Welfare.

If the author's name is not known, books may be located by looking up the title, which is typed at the top of the catalog card above the author's name. Titles of books are usually listed if they are distinctive or if the author is anonymous. Some libraries do not list all the titles beginning History of, Outline of, Report on, etc. Magazines subscribed to by the library are also listed by title.

Filing Rules

The arrangement of the individual catalog cards in the Author-Title Catalog is alphabetical, word-by-word except when a chronological or numerical arrangement is used for clarity. Punctuation marks are disregarded in filing. Examples:

New republic
New York (City) Zoological park
New York times
New yorker
Newsweek

Initial articles: a, an, the, and equivalent terms in other languages, when they appear at the beginning of titles, are disregarded in filing.

Titles and compound names follow single personal surnames. Examples:

Smith, Zachary. (single surname)
Smith college journal. (title)
Smith-Jones, Ivy. (compound surname)
Smith of the Gazette. (title)

Mc, M' and Mac: names beginning with Mc or M' are interfiled as if spelled Mac. Examples:

McCabe, Thomas
Machiavelli, Nicolo
M'Mahon, George.

Umlauts: ä, ë, ü are spelled ae, oe, and ue.

Abbreviations and numerals in general are filed as if spelled out. Examples:

XV	fifteen
400	four hundred

Dr.	Doctor
Mr.	Mister
St.	Saint
U. S.	United States
U. S. S. R.	United Soviet Socialist Republics.

Single letters or initials are filed before a word beginning with the same letter. Examples:

ABC of science.
A. M. Schwartz lectures in physics.
Aarron, Paul.

For a more detailed explanation, look up Seely, Pauline A. A. L. A. rules for filing catalog cards (2nd ed., Chicago, A. L. A., 1968.)

Subject Heading Catalog

Set up a list of specific subject headings related to your subject and revise as the literature search progresses. Look for your subject in Subject Headings Used in the dictionary catalogs of the Library of Congress (7th ed., Washington, D. C., Library of Congress, 1966). Since different individuals may think of the same concept or idea, but use different words to express themselves, standardized subject headings will help you to find out what a library has in its book collection. Subject headings help: 1) show where to find books on any specific subject in the card catalog; 2) show as nearly as possible in one place everything that is to be had on that subject; and 3) indicate through cross-references places to look which may not be immediately apparent.

The Library of Congress Subject Headings is your guide to the subject headings used to index the book collection. If the library has books covering a particular subject, their catalog cards will be found filed in the library's Main Card Catalog behind the appropriate subject heading card. If you are looking for several books on a specific topic, or about a particular person, you need to check for cards which have the subject headings typed at the top of the card above the author's name.

How to find appropriate subject headings

a. Usually the best approach is to be as specific as possible when looking for a subject. For example, Earth as planet has only one class number QB631, while Planets has a range of class numbers from QB601 to QB701. All the books that have the class number QB631 would be about the Earth as a planet, while books on Planets between QB601 and QB701 would have a section, chapter or only paragraphs about the Earth. Or Biomathematics, QH324, would be used instead of Biology, QH301 to QH705.

If a specific subject such as "mercury poisoning" is not found in the Library of Congress Subject Headings, try a larger subject, "water pollution," which would include your subject. If the larger subject is not found, refer to a dictionary for related words that include the subject heading sought. Examples: Earth as planet (QB631) is included in the subject Solar system (QB501 to QB518), which in turn is included under Astronomy (QB). Or Cell nuclei (QB595) is a subdivision of Cells (QH581 to QH671) which is included under Biology (QH301 to QH705).

b. Related or "see also" subject headings are valuable because all information on a subject is seldom found under only one subject heading. Cross references lead you from words or names not used (indicated by an x) to subject headings which are used. They also lead to related subject headings under which more information can be found (indicated by sa or xx). The Library of Congress Subject Headings book lists the many cross references used in the Subject Heading section of the library's card catalog: x, xx or sa. Examples:

Cellulose (Bacteriology, QR160; Botany, QK898.C35)
(Chemistry, QD321; Technology, TS1145)
sa (Cellophane TP986.C4)
(Cellulose industry TS932)
(Hemicellulose QK865)
(Nitrocellulose TP276)
(Pyroxlin TP939)
(Rayon TS1688)
(Wood--Chemistry TA421)
xx (Polysacchrides QD321)

c. If a particular subject heading is not used, try

a synonym. Books with similar content will have the same subject headings, but one book may have more than one subject heading. Since many subjects may be looked for under synonyms or under another form of the subject heading, the "See" references direct the reader to the appropriate subject headings in the dictionary catalog. Examples:

> Chemical geology, See Geochemistry
> Chemical works, See Chemical plants
> Chemiculture, See Hydroponics
> Chemistry of food, See Food--Analysis

d. Subject headings may be inverted. Examples:

> Physics, Astronomical -- See Astrophysics
> Physics, Biological -- See Biological physics
> Physics, Nuclear -- See Nuclear physics
> Physics, Terrestrial -- See Geophysics

e. Complex subjects are usually subdivided. Example:

Biology
- -- Abstracting and indexing (QH301)
- -- Classification (QH83)
- -- Juvenile literature (QH48)
- -- Laboratory manuals (QH324)
- -- Philosophy (QH331)
- -- Tables, etc. (QH310)

Using the Library of Congress Subject Headings book as a guide to what might be found about "Nurses and Nursing" in the Subject Heading Catalog, the following sample list of subject heading and Library of Congress subject classification alpha-numbers was developed by: 1) locating the subject heading "Nurses and Nursing"; 2) finding the class letters RT plus a comprehensive list of cross references to other subject headings; 3) tracing each heading marked x, xx, or sa alphabetically, until a subject class letter or number was eventually found for each cross referenced heading.

NURSES AND NURSING SUBJECT HEADING LIST

Subject Heading	Class Number(s)
Alcoholics---Hospitals and asylums, See Alcoholism-Treatment, Disease	RC565
Almshouses	HV61, HV85-527
Army	UH490-495
Cancer Nursing	RC266
Care of the Sick---Here are books on various types of care, eg. intensive, intermediate, long term and home care including architectural planning, equipment, etc.	
Children---Care and hygiene	RJ61; RJ101
Children---Hospitals	RJ27-28
Clinics---Here are entered works dealing with institutions for treatment of non-resident patients	
Convalescence---(Amusements)	GV1231
Cookery for the sick	RM219
Deaconesses	BV4423-4425
Diet in disease	RM214-259
Dietiticans	RM214-218
Disaster nursing	RT108
Dispensaries	RA960-993
Epileptics---Hospitals, See Charities, Medical Hospitals	RA960-996
First aid in illness and injury	RC86-88
Geriatric nursing	RC954
Home nursing	RT61
Homeopathy---Hospitals and dispensaries	RX6
Hospital administrators	RA972.5
Hospital housekeeping	RA975.5.H6
Hospital libraries	Z675.H7
Hospital pharmacies	RA975.5.P5
Hospitals schools, See Children---Hospitals	
Hospital ships (Naval medicine)	VG450
Hospitalers	BX2825
Hospitals	RA960-996
Hospitals--Accounting	HF5686
Hospitals--Accreditation	RA971
Hospitals--Construction	RA967
Hospitals--Fires and fire prevention	TH9445.H7
Hospitals--Food service	RA975.5.D5
Hospitals--Furniture, equipment, etc.	RA968

Subject Heading	Call Number
Hospitals--Heating and ventilation	RA969
Hospitals--Hygiene	RA969
Hospitals--Inspection	RA971
Hospitals--Juvenile literature	RA965
Hospitals--Laws and legislation	RA974
Hospitals--Location	RA967.7
Hospitals--Outpatient services	RA974
Hospitals--Specifications	RA967.5
Hospitals--Staff	RA972
Hospitals, Convalescent	RA973
Hospitals, Gynecologic and obstetric	RG12-16
Hospitals, Medieval	RA964
Hospitals, Military	UH460-485
Hospitals, Naval & Marine (Marine)	RA975, RA980-993
(Naval)	VG410-450
Hospitals---Nurses, See Hospitals--Staff	RA972
Hospitals, Ophthalmic and aural	RE3
Ear, nose and throat	RF6
Hospitals, Rural	RA975
Hygiene--- (Domestic)	RA771
(Ethical aspects)	BJ1695
(Personal)	RA773-790
Hygiene---Early works to 1800	RA775
Hygiene---Hindu	RA529; RA776
Hygiene---Jewish	RA561
Hygiene---Juvenile literature	QP36-38
Hygiene, Maya	F1435.M4
Hygiene, Public---Congresses	RA422
Hygiene, Public---Exhibitions	RA438
Hygiene, Rural	RA427; RA771
Hygiene, Sexual	HQ31-58
(Men)	RC881
(Women)	RG121
Hygiene---Terminology	RA423
Incurables--Hospitals and asylums	HV3000-3003
Industrial nursing	RC966
Infants--Care and hygiene	RJ61; RJ101
Infirmaries, See Hospitals	
Invalids---Recreation	GV1231
Labor and laboring classes--Medical care	RC963
Labor and laboring classes--Medical examinations	HD7261
Medical centers, See Hospitals	
Medicine	R
Medicine--Biography (Collected)	R134
(By country)	R153-684
Medicine, Clinical	RC61-69

Medicine---Practice	RC
Military Religious orders	CR4701-4775
Navy	VG350-355
Nervous system--Diseases--Hospitals	RC328
Nose--Diseases--Hospitals	RF6
Nurses aides, See Nurses and nursing	
Nurses and nursing	RT
Nurses and nursing--Directories	RT25
Nurses and nursing--Examinations, questions, etc.	RT55
Nurses and nursing--Legal status, laws, etc.	RT4-17
Nurses and nursing--Study and teaching	RT71-81
Nursing as a profession, See Nurses and nursing	
Nursing ethics	RT85
Nursing homes, See Convalescence	
Nursing school libraries, See Libraries, nursing school	Z675.N8
Nursing schools	RT71-81
Obstetrical nursing	RG951
Operating rooms	RA975.5.06
Ophthalmic nursing	RE88
Orthopedia--Hospitals and institutions	RD705
Orthopedic nursing	RD737
Pediatric nursing, See Children--Care and hygiene	
Practical nursing	RT62
Psychiatric Hospitals, See Asylums	
Psychiatric nursing	RC440
Public health nursing,	
See Hygiene, Public	RA
Comprehensive treatises	RA425
General documents	RA11-388
Red Cross	HV560-583
(Army)	UH535-537
(European war)	D628
(Navy)	VG457
School nurses	LB3407
Sick (Aid for,)	HV688-694
Sisters of Charity	BX4452-4470
Surgical nursing	RD99
Therapeutics	RM
Throat--Diseases--Hospitals	RE3; RF6
Trained nurses, See Nurses and nursing	
Tuberculosis nursing	RC311.8
Urological nursing	RC874.7
Volunteer workers in hospitals, See Hospitals	

Use of Catalog Cards

When you have located in the Subject Heading Catalog the card for the book wanted, it will give the following information:

1. Call number, a group of letters and numerals making up the classification number and book number used to locate and identify the book.
2. Author's name, and date of birth and death if it has occurred.
3. Title of the book.
4. Place of publication.
5. Publisher.
6. Date of publication. This is important when the most recent information is needed, or a particular edition is required.
7. Paging, illustrations, height.
8. Notes of interest about the book.
9. Subject headings. They define the book's contents and are headings which should be checked for further information on the same topic.
10. Added entries.

Copy the author's name, the title, and the call number in full exactly as it appears. The call number is found in the upper left corner of the catalog card and is composed of several lines of letters and numbers. If a mistake is made in copying the call number, the book probably will not be found on the shelves.

What's Where?

First, in the Library of Congress Classification System, call numbers begin with the initial capital letters dividing knowledge into broad categories. Each category or LC class is given a different letter. A second letter, followed by numbers, further subclassifies the subject of the book. Originally developed for the books in the Library of Congress, this method of book classification has since been adopted by many university and college libraries. The Library of Congress Classification scheme for science was developed in the early 1900's and was tested before publication in 1905 against the science books in the Library of Congress. The Library of Congress Classification differs notably from Dewey Decimal

Classification in two respects: 1) it is not a decimal system, and 2) the whole field of knowledge is divided into 24 groups using the letters of the alphabet to represent the classes. Provision is made in the Library of Congress Classification for very minute subdivisions of a subject and for expansion.

Classes A to Z:

A GENERAL WORKS. POLYGRAPHY

AE	Encyclopedias (General)
AN	Newspapers
AP	Periodicals (General)

B PHILOSOPHY. PSYCHOLOGY. RELIGION

B-BD) BH-BJ)	Philosophy
BF	Psychology
BL-BX	Religion

C - F HISTORY

DA	Great Britain
DE	Greco-Roman world
DK	Russia
DS	Asia
DT	Africa
E-F	AMERICAS

G - H SOCIAL SCIENCES

G-GF	Geography
GN-GR	Anthropology & Folklore
GV	Recreation
HB-HD, HJ	Economics
HE-HG	Business
HM-HX	Sociology

J - K POLITICAL SCIENCE

JK	United States
JN	Europe
JS	Local government
K	Law

L EDUCATION

LB	Theory and practice
LC	Special aspects

M MUSIC AND BOOKS ON MUSIC

N FINE ARTS
- NA Architecture
- NB Sculpture
- NC Drawing, Design, Illustration
- ND Painting

P LANGUAGE AND LITERATURE

Q SCIENCE (General)
- 300-380 Cybernetics. Information theory
- QA Mathematics
  - 8-10 Mathematical logic
  - 76 Computer science. Electronic data processing
  - 101-145 Elementary mathematics. Arithmetic
  - 152-299 Algebra
    Including machine theory, game theory, probability, mathematical statistics
  - 300-433 Analysis
    Including analytical methods connected with physical problems
  - 443-699 Geometry
  - 611-614 Topology
  - 801-935 Analytic mechanics
    For non-theoretical mechanics, see QC120-168
- QB Astronomy
  - 145-237 Practical and spherical astronomy
  - 275-343 Geodesy
  - 351-421 Theoretical astronomy and celestial mechanics
    Including perturbations, tides
  - 460-991 Astrophysics and descriptive astronomy
    Including stellar spectroscopy, cosmogony
- QC Physics
  - 81-119 Weights and measures
  - 120-168 Descriptive and experimental mechanics
  - 171-197 Constitution and properties of matter. Atomic physics. Including quantum theory, solid-state physics
  - 220-246 Sound. Acoustics
  - 251-338 Heat

	350-495	Light. Optics Including spectroscopy, radiation (General)
	501-768	Electricity and magnetism
	770-798	Nuclear and particle physics. Atomic energy. Radioactivity
	801-809	Geophysics. Cosmic physics
	811-849	Geomagnetism
	851-999	Meteorology
QD		Chemistry
	24-26	Alchemy
	71-142	Analytical chemistry
	146-199	Inorganic chemistry
	241-449	Organic chemistry
	450-655	Physical and theoretical chemistry Including quantum chemistry, stereo-chemistry, chemical reactions, surface chemistry, thermochemistry, solution chemistry, electrochemistry, radiochemistry, photochemistry
	901-999	Crystallography
QE		Geology
	351-399	Mineralogy
	420-499	Petrology
	500-625	Dynamic and structural geology
	650-699	Stratigraphic geology
	701-996	Paleontology Including paleozoology and paleobotany
QH		Natural history
	75-77	Nature conservation. Landscape protection
	201-278	Microscopy
	301-559	General biology Including life, genetics, evolution, reproduction, ecology
	573-671	Cytology
QK		Botany
	641-707	Morphology, anatomy, embryology, and histology
	710-899	Plant physiology
	901-987	Ecology
QL		Zoology
	750-797	Animal behavior and pyschology
	801-950	Anatomy
	951-991	Embryology

QM		Human anatomy
	601-699	Human embryology
QP		Physiology
	351-499	Physiological psychology
	501-801	Animal biochemistry
	901-981	Experimental pharmacology
QR		Microbiology
	75- 99	Bacteria
	180-189	Immunology
	355-484	Virology

R MEDICINE

S AGRICULTURE

T TECHNOLOGY

U - V MILITARY & NAVAL SCIENCES

Z BIBLIOGRAPHY AND LIBRARY SCIENCE

The Library of Congress Classification Schedules, published as separate volumes, each with its own index, can be of help in defining subdivisions of science and technology and making a clear delineation of information sought. In general, scientific and technological subjects are assigned to the following Library of Congress letters:

Agriculture	S
Astronomy	QB
Bibliography	Z
Botany	QK
Chemistry	QD
Geography	G-GF
Geology	QE
Human Anatomy	QM
Mathematics	QA
Medicine	R
Microbiology	QR
Military Science	U
Natural History	QH
Naval Science	V
Physics	QC
Physiology	QP
Science (General)	Q
Techology	T
Zoology	QL

Under the Dewey Decimal System, books are distributed by classification or general subjects as follows:

000 -- 099	General Works
100 -- 199	Philosophy and Psychology
200 -- 299	Religion
300 -- 399	Social Science
400 -- 499	Languages
500 -- 599	Pure Science
600 -- 699	Applied Science & Technology
700 -- 799	Fine Arts
800 -- 899	Literature
900 -- 999	History, Biography, Geography.

Each of the 10 classes is further divided into the main subdivisions of the subject. For example, 500 represents Pure Science and the subdivisions in this class are:

500	Pure Science
510	Mathematics
520	Astronomy & Allied Sciences
530	Physics
540	Chemistry
548	Crystallography
549	Mineralogy
550	Earth Sciences
560	Paleontology
570	Anthropology & Biology
580	Botanical Sciences
590	Zoological Sciences

Most scientific and technical knowledge is classified between the numbers 500 and 700. A detailed index in a separate volume makes it possible to locate the specific Dewey classification number in the classification tables.

A book will have a call number on its spine or front identical to the call number on its card in the card catalog. Books are located on the shelves according to the letters and numbers that make up the call number. The call number is the book's Library of Congress or Dewey Decimal subject classification number plus the book number which codes author, title and number of volume or copies. First, in alphabetical order, the Library of Congress classification letter:

QA QB QC R S.

Then, in numerical order by the second line, the classification number:

QA	QA	QC	R	RC
76	300	81	735	560.

Then, in alphabetical and numerical order by the third line, the book number, which is treated as the first decimal place:

QA	QA	QA	QA	QA
76	76	76	76	76
.A8	.A82	.C6	.D32	.D4.

If the fourth line begins with a letter, not a number, it also is treated as a decimal. If there is a decimal point anywhere in the call number, all numbers following on the same line are to be treated as decimals. Sometimes the call number is four or five lines. It should always be considered line by line until the book you are seeking is located.

The same procedure is used to locate a book classified according to Dewey Decimal numbers. Any number after the decimal point is considered a decimal. Even though there is no decimal point before the class number, it too is considered a decimal, for example:

511.08	521	521.183	521.4	521.4
D61a	R52c	B31d	A451n	A46m.

Assuming that you now have a list of call numbers, either Dewey Decimal or Library of Congress, the next step is to physically locate subject related books and periodicals on the shelf in the library's collections by call number.

Remember to look at every book that has the same subject classification number, because you may find more comprehensive or recent information on your subject than that what you had sought originally. Examine each book's table of contents and index, then select those that seem subject-related for further reading. Make a bibliographic record of each book selected including author, title, place, publisher, date, edition, and your personal comments about the book's contents. Whenever you examine any book that is subject-related, including reference books, make a point to record its call number. Use your lists of classification numbers and subject headings as a bibliographic guide for locating additional book call numbers and periodical titles that may relate to your subject.

Other References

There are other book listings referred to by librarians which supplement the Main Card Catalog. The Shelf List or Location File is one. Serials Record is another. Neither is arranged alphabetically, but in the same call number order as the books are arranged on the shelves. Both listings are a record of the books and periodical volumes in a particular library collection. They both serve as sources of additional bibliography on specific subjects.

Chapter IV

SPECIAL REFERENCE WORKS

Enlarge the scope of the search as necessary by adding new periodical titles; consult periodical articles indicated by the clues that appear as the library search progresses in indexes, abstracts and reviews.

The most recent and often most detailed information and bibliography on a specific subject is usually found in a currently published periodical article. It's important to remember that articles from magazines or periodicals are not listed in the Main Card Catalog. First, check the bound volume of the periodical for the annual index, then search in one of the periodical indexes to obtain the exact reference. You can find whether or not a particular publication is indexed or abstracted, and if so where, by looking up that periodical in *Ulrich's Periodicals Directory*. Indexes and abstracts may be of periodicals, publications of learned societies, documents, book reviews, or composite books. To the student searching the library the importance of a good index, whether it be to an individual book, a separate periodical, or a group of periodicals or books, cannot be over-estimated. Indexes of various kinds are found in all library reference collections. The compilation of subject or author bibliographies and the verification of bibliographic details could not be done without these versatile, comprehensive publications. The value of the wide range of subject matter currently available in periodicals cannot be overstressed. A thorough study of this source of information, written for both the layman and the expert, gives the student opportunities for gaining fresh data frequently unobtainable from other sources.

The term "periodicals," often called "serials" by the librarian, includes magazines, journals and other similar publications. They range from general titles such as *Scientific American* and *Science* to definitive or scholarly journals. As the weekly, monthly or quarterly issues accumulate, the periodicals are usually bound into permanent volumes and intershelved with books of the same classification in the science collection, or are made available reproduced on microfilm.

In order to locate information in periodicals on a certain subject or by a particular author from current and past issues, it is necessary to search in one of the many periodical indexes. The first step in starting a search through the general and specialized periodical indexes is to find out which is likely to include the subject information. Indexes dealing with special information are useful to those desiring extensive, detailed, technical and specific data. Confined to a special subject, an index of this type covers sources of written material often of limited circulation, or sources of highly specialized articles published for small groups of readers.

Abstracts and reviews are similar to periodical indexes. In addition to the usual bibliographic citation, a review or a summary or abstract of the article, book, or document is given. The specialist who lacks the time to cover all the current material in his field finds the most important articles summarized for him in the abstracts; he can then choose which articles he wishes to read in their entirety. Abstracting services carefully endeavor to publish adequate and accurate abstracts of the most important current articles in periodicals and society publications, doctoral dissertations, books, and technical papers containing new information of subject interest. Abstracts are also useful as reference sources and as guides to reference material. The comments and opinions expressed in abstracts or reviews of books and other publications are usually those of the individual reviewers and not of the editor or publisher. If the abstracts and reviews of books are checked monthly, many of the new and/or important titles will be uncovered. Unfortunately, there is sometimes a long delay before a new book is announced in the abstracting journals. Some abstracts report new technological information revealed in the patent literature. Abstracts are sometimes available in printed form, on microfilm or on computer readable tape. An abstract should annotate all essential information, including negative results, be brief as possible, and yet accurately reflect the published article. Abstracts and reviews are generally more restricted in their subject coverage than the periodical indexes.

The scope of the periodical search is increased materially by the use of back files of abstracts and other reviewing journals because they cover and include obscure and fringe interest publications in which occasional articles of importance appear. Obviously no one library can afford to subscribe to all the journals and periodicals published and must depend upon abstracting services for information on what is currently published about a subject.

Selection of subject headings, key words or descriptors for searching indexes and abstracts must fit not merely the requirements of the library search but also the specific periodical index searched.

For example, many of the terms listed in the Library of Congress Subject Headings, which is the authority for all libraries using L.C. printed cards in their catalogs, will not be suitable for Applied Science and Technology Index or Engineering Index. The search must be translated into the exact nomenclature or language of each index, abstract, and review. This means the searcher should become familiar with the vocabulary of the subject searched as well as the indexing practices of the various indexing and abstracting services. To be on the safe side, the library searcher should look up every pertinent subject heading, from the general to the specific, that is cross referenced directly or implied during the search. Since the various indexes, abstracts and reviews overlap in coverage of different subject areas, each of the following titles contains a list of periodicals and subject headings related to the topics indexed.

Indexes to Periodicals

GENERAL SCIENCE

American documentation. Washington, D.C., American Society for Information Science, v. 1-- 1950--

Current contents, life sciences. (Formerly: Current contents, chemical, pharmaco-medical & life science edition.) Philadelphia, Institute for Scientific Information, v. 1-- 1958--

Current contents, physical sciences. (Formerly: Current contents of space, electronic and physical sciences.)

Philadelphia, Institute for Scientific Information, v. 1-- 1961--

Current index to conference papers in life sciences. New York, C. C. M. Information Sciences, v. 1-- 1969--

Information science abstracts. (Formerly: Documentation abstracts.) (American Society for Information Science Special Libraries Assn. & American Chemical Society, Division of Chemical Literature.) Philadelphia, Documentation Abstracts, Inc., v. 1-- 1966--

International Federation for Documentation. Abstracting services in science, technology, medicine, social sciences, humanities. The Hague, 1965.

News from science abstracting & indexing services. Philadelphia, National Federation of Science Abstracting and Indexing Services, v. 1-- 1958--

Pandex current index of scientific and technical literature. (Separate author subject sections.) New York, C. C. M. Information Sciences Inc., v. 1-- 1969--

Proceedings in print. Mattapan, Mass., Proceedings in Print, Inc., v. 1-- 1964--

Readers' guide to periodical literature. Bronx, N. Y., H. W. Wilson Co., v. 1-- 1900--

Science citation index. Philadelphia, Institute for Scientific Information, v. 1-- 1963--. Permuterm Subject Index, 1967--

Subject index to United States joint publications. Washington, D. C., Research and Microfilm Publcations, Inc., v. 1-- 1966--

Vertical file index; a subject and title index to selected pamphlet material. Bronx, N. Y., H. W. Wilson Co., v. 1-- 1932--

Newspapers print the facts about an important scientific discovery the day it is announced. The date of an event is the clue needed and an index of dates, or an index to one newspaper, will furnish a workable index to all newspapers for subjects of general interest. Several newspaper indexes accurately cover pratically all current events worthy of note, including a great many in the world of science with thorough cross references. These are

New York Times Index. Semimonthly, cumulated annually.
Wall Street Journal Index. Monthly, cumulated annually.

MATHEMATICS

Computing reviews. New York, Association for Computing Machinery, v. 1-- 1960--

Contents of contemporary mathematical journals. Providence, R.I., American Mathematical Society, v. 1-- 1969--

International abstracts in operations research. Baltimore, Md., Operations Research Society, v. 1-- 1961--

Mathematical reviews. Providence, R.I., American Mathematical Society, v. 1-- 1940--

PHYSICS

Cumulative solid state abstracts. Cambridge, Mass., Cambridge Communications Corp., v. 1-- 1957--

Index to the literature of magnetism. New York, American Institute of Physics, v. 1-- 1961--

Nuclear magnetic resonance abstracts service. Evanston, Ill., Preston Technical Abstracts Co., v. 1-- 1964--

Nucleic acids monthly. Kettering, Ohio, Literature Searchers, v. 1-- 1965--

Science abstracts. Section A: Physics abstracts. London, Eng., Institution of Electrical Engineers and New York, Institute of Electrical & Electronics Engineers, Inc., v. 1-- 1898--

CHEMISTRY

Accounts of chemical research. Washington, D. C., American Chemical Society, v. 1-- 1968--

Chemical abstracts. Washington, D. C., American Chemical Society, v. 1-- 1907--
  Chemical abstracts - Applied chemistry sections, v. 1-- 1963--
  Chemical abstracts - Macromolecular sections, v. 1-- 1963--
  Chemical abstracts - Organic chemistry sections, v. 1-- 1963--
  Chemical abstracts - Physical chemistry sections, v. 1-- 1963--
  Chemical abstracts - Biochemistry sections, v. 1-- 1963--

Chemical market abstracts. New York, Foster D. Snell Inc., v. 1-- 1950--

Chemical titles. Washington, D. C., American Chemical Society, v. 1-- 1961--

Chemical-biological activities (CBAC). Washington, D. C., American Chemical Society, v. 1-- 1965--

Howell, M. G., et al, eds. Formula index to NMR literature data. New York, Plenum, v. 1-- 1965--

Index chemicus. Philadelphia, Institute for Scientific Information, v. 1-- 1960-- (Cumulated annually as Encyclopaedia Chimica Internationalis.)

Information services from Chemical Abstract Service. Washington, D. C., American Chemical Society, v. 1-- 1970--

Intra-science chemistry reports. Santa Monica, Cal., Intra-Science Research Foundation, v. 1-- 1967--

Plastic industry notes. (Supersedes: Polymer and plastics business abstracts.) Columbus, Ohio, The Ohio State University, v. 1-- 1967--

Polymer industry news. Akron, Ohio, Center for Information Systems, Univ. of Akron, v. 1-- 1967--

Polymer literature abstracts. (Formerly: Current awareness bulletin.) Akron, Ohio, Center for Information Systems, Univ. of Akron, v. 1-- 1965--

Polymer science & technology (Post). Washington, D.C., American Chemical Society, v. 1-- 1967--. Issued in two sections: POST-J and POST-P.

Selenium & tellurium abstracts. Columbus, Ohio, Chemical Abstracts Service, v. 1-- 1959--

EARTH SCIENCES

Abstracts of papers. International Union of Geodesy and Geophysics, v. 1-- 1963--

Corbin, J. B., comp. An Index of state geological survey publications issued in series. Metuchen, N.J., Scarecrow, 1965--

Earth science reviews. New York, American Elsevier, v. 1-- 1966--

Fogel, L. J. Composite index to marine science and technology. San Diego, Cal., Alfo Pub., 1966--

Geoscience abstracts. (Formerly Geological abstracts.) Washington, D.C., American Geological Institute, v. 1-- 1953--

International geology review. Washington, D.C., American Geological Institute, v. 1-- 1959--

Meteorological and geoastrophysical abstracts. Boston, American Meteorological Society, v. 1-- 1950--

Oceanic citation journal. (Supersedes: Oceanic abstracts.) (With selected abstracts.) La Jolla, Cal., Oceanic Research Institutes, v. 1-- 1968--

Oceanic index. (Includes: Oceanic citation journal.) La Jolla, Cal., Oceanic Information Center, v. 1-- 1964--

Oceanic instrument reporter. La Jolla, Cal., Ocean Engineering Information Service, v. 1-- 1968--

Water resources abstracts. Urbana, Ill., American Water Resources Assn., v. 1-- 1968--

BIOLOGICAL SCIENCES

Abstracts of mycology. Philadelphia, Biosciences Information Service of Biological Abstracts, v. 1-- 1967--

Bacteriological reviews. Baltimore, Williams & Wilkins, v. 1-- 1947--

Biological abstracts, from the world's biological research literature, v. 1-- 1926--
Issued in the following sections:
A. General biology, 1939-62
B. Basic medical sciences, 1939-62
C. Microbiology, immunology, parasitology, 1939-62
D. Plant sciences, 1939-62
E. Animal sciences, 1939-62
F. Animal production and veterinary science, 1942-53
G. Food and nutrition research, 1943-53
H. Human biology, 1946-53
I. Cereal products, 1947-53.

Includes B. A. S. I. C. (Biological Abstracts Subjects in Context) computer arranged subject index, v. 1-- 1962--

Also included with subscription:
Bioresearch index. v. 1-- 1965--

All published by Philadelphia, Biosciences Information Service of Biological Abstracts.

Biological and agricultural index. Bronx, N. Y., H. W. Wilson Co., v. 50-- 1964-- (Continues Agricultural index, v. 1-49, 1916-64.)

Communications in behavioral biology. (Part A: Original Articles; Part B: Abstracts and Index.) New York, Academic Press, 1970--

Quarterly review of biology. Stony Brook, N. Y., Stony Brook Foundation, v. 1-- 1926--

MEDICAL SCIENCES

Abstracts of current literature on venereal disease. Atlanta, Ga., National Communicable Disease Center, Venereal Disease Branch, v. 1-- 1966--

Abstracts of hospital management studies. Ann Arbor, Mich., Univ. of Michigan, Cooperative Information Center for Hospital Management Studies, v. 1-- 1963-64--

American journal of diseases of children. Chicago, Ill., American Medical Assn., v. 1-- 1911--

American Medical Association. Journal. Chicago, Ill., American Medical Assn., v. 1-- 1848--

Archives of environmental health. (American Academy of Occupational Medicine; American College of Preventive Medicine and Association of Teachers of Preventive Medicine.) Chicago, Ill., American Medical Assn., v. 1-- 1950--

Archives of neurology. (American Neurological Assn.) Chicago, Ill., American Medical Assn., v. 1-- 1959--

Archives of otolaryngology. Chicago, Ill., American Medical Assn., v. 1-- 1925--

Birth defects. New York, National Foundation, v. 1-- 1964--

Blood monthly. Kettering, Ohio, Literature Searchers, v. 1-- 1968--

Brain biochemistry monthly. Kettering, Ohio, Literature Searchers, v. 1-- 1966--

Cancer chemotherapy abstracts. Bethesda, Md., National Cancer Institute, v. 1-- 1960--

Cardiovascular compendium. Philadelphia, Pa., Compendium Publications, v. 1-- 1965--

Circulation. New York, American Heart Assn., v. 1-- 1950--

Clinical research. Washington, D. C., American Federation for Clinical Research, v. 1-- 1953--

Communication disorders. Baltimore, Md., Information Center for Hearing, Speech and Disorders of Human Communication, The Johns Hopkins Medical Institutions, v. 1-- 1968--

Computers in medicine abstracts. New York, CIDCOMED (Council for Interdisciplinary Communications in Medicine), v. 1-- 1968--

Cumulative index to nursing literature. Glendale, Cal., Glendale Adventist Hospital Association, v. 1-- 1956-- Also: Nursing subject headings, 4th ed., 1970.

Current tissue culture literature. New York, October House Inc., v. 1-- 1966--

Cystic fibrosis. New York, National Cystic Fibrosis Research Foundation, v. 1-- 1962--

Dental abstracts. Chicago, American Dental Association, v. 1-- 1956--

Edema. New York, Excerpta Medica Foundation, v. 3-- 1969--

Excerpta Medica. New York, Excerpta Medica Foundation:
- Section 1: Anatomy, Anthropology, Embryology & Histology, v. 1-- 1947--
- Section 2: Physiology, v. 1-- 1948--
- Section 3: Endocrinology, v. 1-- 1947--
- Section 4: Microbiology, Bacteriology, Virology, Mycology and Parasitology (Formerly: Medical Microbiology, Immunology & Serology), v. 1-- 1948--
- Section 5: General Pathology and Pathological Anatomy, v. 1-- 1948--
- Section 6: Internal Medicine, v. 1-- 1947--
- Section 7: Pediatrics, v. 1-- 1947--
- Section 8: Neurology and Neurosurgery, v. 1-- 1948--
- Section 9: Surgery, v. 1-- 1947--
- Section 10: Obstetrics and Gynecology, v. 1-- 1948--
- Section 11: Oto-Rhino-Laryngology, v. 1-- 1948--
- Section 12: Ophthalmology, v. 1-- 1947--
- Section 13: Dermatology and Venereology, v. 1-- 1947--

Section 14: Radiology, v. 1-- 1947--
Section 15: Chest Diseases, Thoracic Surgery and Tuberculosis, v. 1-- 1948--
Section 16: Cancer, v. 1-- 1953--
Section 17: Public Health, Social Medicine & Hygiene, v. 1-- 1955--
Section 18: Cardiovascular Diseases and Cardiovascular Surgery, v. 1-- 1957--
Section 19: Rehabilitation and Physical Medicine, v. 1-- 1958--
Section 20: Gerontology and Geriatrics, v. 1-- 1958--
Section 21: Developmental Biology and Teratology, v. 1-- 1961--
Section 22: Human Genetics, v. 1-- 1963--
Section 23: Nuclear Medicine, v. 1-- 1964--
Section 24: Anesthesiology, v. 1-- 1966--
Section 25: Hematology, v. 1-- 1967--
Section 26: Immunology, Serology and Transplantation, v. 1-- 1967--
Section 27: Biophysics, Bioengineering and Medical Instrumentation, v. 1-- 1967--
Section 28: Urology and Nephrology, v. 1-- 1967--
Section 29: Biochemistry, v. 1-- 1948--
Section 30: Pharmacology and Toxicology, v. 1-- 1948--
Section 31: Arthritis and Rheumatism, v. 1-- 1965--
Section 32: Psychiatry, v. 1-- 1948--
Section 33: Orthopedic Surgery, v. 1-- 1956--.

Geriatrics digest. Northfield, Ill., Geriatrics Digest Inc., v. 1-- 1968--

Hospital abstract service. Berwyn, Ill., Physicians' Record Co., v. 1-- 1934--

Hospital literature index. Chicago, American Hospital Association, v. 1-- 1945--

Index to dental literature. Chicago, American Dental Association, v. 1-- 1921--

International neurosciences abstracts. North Hollywood, Cal., Ed. S. Grossman, v. 1-- 1968--

International nursing index. New York, American Journal of Nursing Co., v. 1-- 1966--

International pharmaceutical abstracts. Washington, D.C.,

American Society of Hospital Pharmacists, v. 1-- 1964--

Journal of laboratory and clinical medicine. (Central Society for Clinical Research.) St. Louis, Mo., C. V. Mosby Co., v. 1-- 1915--

Laser abstracts for the medical profession. Evanston, Ill., Lowry-Cocroft Abstracts, v. 1-- 1966--

Leukemia abstracts. (Lenore Schwartz Leukemia Research Foundation.) Chicago, John Crerar Library, v. 1-- 1953--

Medical abstract service. Berwyn, Ill., Physicians' Record Co., v. 1-- 1943--

Medical care review. (Formerly: Public health economics and medical care abstracts.) Ann Arbor, Mich., University of Michigan, Public Health School, v. 1-- 1944--

Multiple sclerosis abstracts. New York, National Multiple Sclerosis Society, v. 1-- 1956--

Muscular dystrophy abstracts. New York, Excerpta Medica Foundation, v. 1-- 1957--

Oral research abstracts. Chicago, American Dental Assn., v. 1-- 1966--

Pediatrics digest. Northfield, Ill., Pediatrics Digest, Inc., v. 9-- 1967--

Pharmaceutical abstracts. Austin, Texas, Univ. of Texas, College of Pharmacy, v. 9-- 1968--

Pharmacologist. Bethesda, Md., American Society for Pharmacology & Experimental Therapeutics, Inc., v. 1-- 1959--

Physicians basic index. Kettering, Ohio, Charles F. Kettering Memorial Hospital, v. 1-- 1966--

Physiological reviews. Baltimore, American Physiological Society, v. 1-- 1921--

Quarterly journal of studies on alcohol. New Brunswick, N.J., Rutgers Center of Alcohol Studies, v. 1-- 1940--

Surgery, gynecology & obstetrics. (with International abstracts of surgery.) (American College of Surgeons.) Chicago, Martin Memorial Foundation, v. 1-- 1905--

Survey of anesthesiology. Baltimore, Md., Williams & Wilkins, v. 1-- 1957--

Survey of ophthalmology. Baltimore, Md., Williams & Wilkins Co., v. 1-- 1956--

Tissue culture abstracts. Grand Island, N.Y., Grand Island Biological Co., v. 1-- 1963--

Upjohn research abstracts. (Formerly: Upjohn research abstracts & Upjohn clinical abstracts.) New York, Farley Manning Associates, v. 1-- 1957--

Vitamin abstracts. Chicago, Assn. of Vitamin Chemists, v. 1-- 1946--

AGRICULTURAL SCIENCES

Abstracts for the advancement of industrial utilization of wheat. Pullman, Wash., Washington State Univ., College of Engineering, Research Div., v. 1-- 1962--

Gardener's abstracts. Goleta, Cal., Leisure Abstracts, v. 1-- 1969--

Institute of Paper Chemistry. Abstract bulletin. (Keyword Supplement '300 & Computer Tape '1000.) Appleton, Wis., Institute of Paper Chemistry, v. 1-- 1930--

Quarterly cumulative veterinary index. Arvada, Colo., Index Inc., v. 1-- 1966--

Tobacco abstracts. Raleigh, N.C., Tobacco Literature Service, D. H. Hill Library, North Carolina State Univ., v. 1-- 1956--

Turkey world. Mount Morris, Ill., Watt Publishing Co., v. 43-- 1968--

Update/agricultural chemicals. Philadelphia, Information Co. of America, v. 1-- 1969--

ENGINEERING SCIENCES

A P C A abstracts. Section of Journal. Pittsburgh, Air Pollution Control Association, v. 1-- 1955--

A S C E publications abstract. New York, American Society of Civil Engineers, v. 2-- 1967--

Abstract review. Washington, D. C., National Paint, Varnish & Lacquer Assn., v. 1-- 1928--

Abstracts of computer literature. (In 2 Parts: Part 1: Marketing Edition; Part 2: Engineering Edition.) Pasadena, Cal., Burroughs Corp., Business Machines Group, v. 1-- 1957--

Air pollution titles. University Park, Pa., Pennsylvania State University, Center for Air Environment Studies, v. 1-- 1965--

Alphabetical subject index to Petroleum abstracts. Tulsa, Okla., Univ. of Tulsa, Information Services Dept., v. 1-- 1961--

American Ceramic Society. Bulletin. Columbus, Ohio, American Ceramic Society, v. 1-- 1918--

__________. Journal (Ceramic abstracts.) Columbus, Ohio, American Ceramic Society, v. 1-- 1918--

American Petroleum Institute. Abstracts of petroleum substitutes literature and patents. New York, American Petroleum Institute, Central Abstracting and Indexing Service, v. 1-- 1969--

Applied mechanics reviews. Easton, Pa., American Society of Mechanical Engineers, v. 1-- 1948--

Applied science & technology index. New York, H. W. Wilson Co., v. 46-- 1958--. Supercedes in part and continues Industrial arts index, 1913--

Batelle technical review. Columbus, Ohio, Batelle Memorial Institute, v. 1-- 1952--

Boating abstracts. Goleta, Cal., Leisure Abstracts, v. 1-- 1969--

Computer & control abstracts/C C A. New York, Institute of Electrical and Electronic Engineers, Inc., v. 1-- 1969--

Corrosion abstracts (U.S.). Houston, Tex., National Assn. of Corrosion Engineers, v. 1-- 1962--

Current index to conference papers in engineering. New York, C. C. M. Information Sciences, v. 1-- 1969--

Dual dictionary coordinate index to petroleum abstracts. Tulsa, Okla., Univ. of Tulsa, Information Services Dept., v. 1-- 1961--

Electrical & electronics abstracts/E E A. New York, Institute of Electrical and Electronics Engineers, Inc., v. 1-- 1969--

Electronics abstracts journal. Washington, D.C., Cambridge Communications Corporation, v. 1-- 1966--

Engineering index. New York, Engineering Index, Inc., v. 1-- 1884--. Services include: Card-A-Lert (cards covering subject categories); Plastics Monthly; Compendex (Computerized Engineering Index.)

Gas abstracts. Chicago, Institute of Gas Technology, v. 1-- 1945--

Gas chromatography abstracting service. Evanston, Ill., Preston Technical Abstracts Co., v. 1-- 1958--

Home economics research abstracts. Washington, D.C., American Home Economics Assn., v. 1-- 1967--

I S A instrumentation index. Pittsburgh, Instrument Society of America, v. 1-- 1967--

Industrial hygiene digest. Pittsburgh, Industrial Hygiene Foundation, v. 29-- 1965--

International aerospace abstracts. New York, A I A A Technical Information Service, v. 1-- 1961--

Laser abstracts. Evanston, Ill., Lowry-Cocroft Abstracts, v. 1-- 1963--

Metals abstracts (Formed by the merger of: Review of metal literature and Metallurgical abstracts.) Metals Park, Ohio, American Society for Metals, v. 1-- 1968--

Microelectronics abstracts. Evanston, Ill., Lowry-Cocroft Abstracts, v. 1-- 1965--

Petroleum abstracts. Tulsa, Okla., Univ. of Tulsa, Information Services Dept., v. 1-- 1961--

Plastics monthly. New York, Engineering Index Inc., v. 5-- 1969--. Supercedes Engineering Index: Plastics Section, v. 1-4, 1965-68.

Pollution abstracts. La Jolla, Cal., Oceanic Library and Information Center, v. 1-- 1970--

Quality control and applied statistics. Whippany, N.J., Executive Sciences Institute, Inc., v. 1-- 1956--

Science abstracts. Section B: Electrical & electronics abstracts. London, Institution of Electrical Engineers and New York, Institute of Electrical & Electronics Engineers, Inc., v. 1-- 1898--

Science abstracts. Section C: Computer & control abstracts. London, Institution of Electrical Engineers and New York, Institute of Electrical & Electronics Engineers, Inc., v. 1-- 1966--

Search. Fort Lee, N.J., Compendium Publishers Int'l. Corp., v. 1-- 1964--
Search: Chemical materials & products division.
Search: Coal, coke & mineral tars division.
Search: Drugs division.
Search: Dyes, pigments & coatings division.
Search: Essential oils, soaps & toiletries division.
Search: Fertilizers division.
Search: Foodstuffs division.
Search: Inorganic chemicals division.
Search: Metals division.
Search: Non-metallic minerals division.
Search: Oils, fats & waxes division.
Search: Organic chemicals division.
Search: Pesticides division.
Search: Petroleum division.
Search: Plastics & resins division.

Search: Pulp & paper division.
Search: Rubber division.
Search: Textiles division.

Selected Rand abstracts. Santa Monica, Cal., Rand Corp., Reports Dept., v. 1-- 1963--

Solar energy: the journal of solar energy science and engineering. Section: solar abstracts. Phoenix, Ariz., Assn. for Applied Solar Energy, v. 1-- 1957--

Solid rocket structural integrity abstracts. Salt Lake City, University of Utah, College of Engineering, v. 1-- 1965--

Textile technology digest. Charlottesville, Va., Institute of Textile Technology, v. 1-- 1944--

Cross References

Sometimes when you are looking for material on a particular subject you find no articles under the heading you check but you are instructed to look under another heading. Or you may find a subject heading which has a number of periodical articles listed under it and after the last article are directions to look under some other headings also. This means that there are articles under these other headings which might give additional information. Add these new headings to your list as they are uncovered and search accordingly.

Examples of subject headings and cross references used in Biological Abstracts:

INTEGUMENTARY SYSTEM
  Anatomy
  General; Methods
  Pathology
  Physiology and Biochemistry

INVERTEBRATE MORPHOLOGY, PHYSIOLOGY AND PATHOLOGY, EXPERIMENTAL AND COMPARATIVE
  Acanthocephala
  Annelida
  Arthropoda
    Arachnoidea
    Crustacea

General
Insecta
General
Morphology, Comparative
Pathology
Physiology
Myriapoda
Aschelminthes
Brachiopoda
Chaetognatha
Ctenophora
Echinodermata
Echiuroidea
Ectoprocta
Entoprocta
General
Hemichordata
Linguatulida
Mesozoa
Mollusca
Onychophora
Phoronidea
Platyhelminthes
Pogonophora
Porifera
Protozoa
Rhynchocoela
Sipunculoidea
Tardigrada

IN-VITRO STUDIES (Cellular and Sub-cellular)

JOINTS (See Bones, Joints, Fasciae, Connective and Adipose Tissue)

LABORATORY ANIMALS
General
Gnotobiology

LYMPHATIC SYSTEM (see Blood, Blood-Forming Organs and Body Fluids)

MATHEMATICAL BIOLOGY AND STATISTICAL METHODS

MEDICAL MICROBIOLOGY (includes Veterinary)
Bacteriology
General; Methods

Mycology
Phycology
Virology

MEDIA, TISSUE CULTURE (see Tissue Culture, Apparatus, Methods and Media)

METABOLISM
Carbohydrates
Disorders
Energy and Respiratory Metabolism
General Metabolism; Metabolic Pathways
Lipids
Minerals
Nucleic Acids, Purines and Pyrimidines
Plant (see Plant Physiology, Biochemistry and and Biophysics)
Porphyrins and Bile Pigments
Proteins, Peptides and Amino Acids
Sterols and Steroids
Vitamins
Fat-Soluble Vitamins
General
Water-Soluble Vitamins

MICROBIOLOGY
Clinical Methods (see Clinical Microbiological Methods (includes Veterinary))
Food and Industrial (see Food and Industrial Microbiology)
Medical (see Medical Microbiology (includes Veterinary))
Public Health (see Public Health Microbiology)
Soil (see Soil Microbiology)
Taxonomy (see General and Systematic Bacteriology; Virology, General)
Veterinary (see Veterinary Science)

Periodical indexes may use "see also" cross references to related subject headings. An example of "see also" references from the Applied Science and Technology Index:

Astronomy

See also
Comets

Eclipses
Galactic systems
Jupiter (planet)
Life on other planets
Mars (planet)
Meteors
Moon
Orbits
Planets
Radio astronomy
Stars
Venus (planet)

Add whatever new "see" and "see also" references are found. Some indexes, however, do not supply "see also" references. Some abstracts may use a keyword (Key-Word-In-Context, KWIC) subject index to provide a quick entry into the subject content of the abstracts. One or more keyword entries are derived for each abstract from the title, text, or context of the abstract. Synonyms are not usually included as additional keywords for the same abstract; no major effort has been directed toward standardization of synonyms. Abbreviations and acronyms employed in keyword phrases are sometimes standardized. A limited number of cross-references are included to indicate other keywords or phrases which should be examined. Use your dictionary to identify related terms, then check other subject related abstracts and reviews for articles and other printed sources of information.

Indexed articles are usually listed by subject or by author and title. In some indexes, every author's name appearing on the original article is listed alphabetically in the author index, including corporate authors for organizational or society reports. Most give the following data: title of article, author, periodical title, volume number, inclusive paging, and date of issue.

Some indexes state whether the article is illustrated or has diagrams and maps. Keys to these abbreviations usually appear in the front of the index or abstract volume. Also listed alphabetically are the periodical titles that are regularly indexed. Usually, the information is given in abbreviated form. Following is an example of a typical indexed citation:

INTERNAL combustion engines
Testing
Telemetering of information from a working internal-combustion engine. M. H. Westbrook and R. Munro. bibliog il diags J. Eng Power 89:247-54 Ap '67.

This sample entry means that M. H. Westbrook and R. Munro wrote an article entitled, "Telemetering of information from a working internal-combustion engine." The article appeared in the April 1967 issue of the Journal of Engineering for Power on pages 247 to 254. It contains illustrations, diagrams and a bibliography.

When a periodical article related to your subject is found in the index, it is important to copy the complete reference: title of the magazine, volume number, inclusive pages, and the date of the issue, to aid in locating the actual article.

For topics or data on any subject, it helps to begin with the most recent paperbound number of the index and to work back chronologically through the clothbound volumes until the desired year is reached. Keep in mind that subject headings may change from year to year within a set of indexes. During the current year the abstracts and indexes for each volume appear in various monthly or quarterly paperbound installments. These installments are collected at the year's end to make up the annual volume. In order to follow developments in a subject, the library searcher is advised to check his subject headings monthly in the appropriate indexing or abstracting journals. The current issues may provide only subject heading indexes but they help indicate what new research progress is being made.

How to Locate a Periodical Article

No general statement could possibly apply to the physical layout of every type of library. However, there are usually one or more subject reference areas, depending on the size of the library. Here are shelved dictionaries, encyclopedias, handbooks, indexes, and current periodicals most often used. Current issues of all periodicals may be displayed in a separate reading area. The usual practice of most academic librarians is to shelve back issues and bound volumes of periodicals in the stacks, together with the books on the same subject, arranged by their call numbers according to the Dewey Decimal or the Library of Congress classification.

Students preparing to conduct library research should understand that in shelving periodicals, policy may differ among libraries even though the same Dewey or Library of Congress classification system is used. Special location shelving arrangements may be made for different types of publications such as atlases, maps, theses, foreign language materials and government documents. Folio and larger sized periodicals, regardless of their information content, may be shelved together.

A given volume and number of a specific periodical may be positively identified by looking up the call number in the library's Main Card Catalog and the periodical then obtained by showing this call number to the reference librarian. If a number of periodicals must be searched, the student should learn the library's physical layout, the Dewey and/or Library of Congress classification schedule (some libraries use both) for science and technology, and the library's policy and regulations for borrowing, loan periods, etc.

For ease in determining a library's periodical holdings in bound volumes, some librarians file a catalog card for each periodical in a separate section of the card catalog called a Serial Record, arranged by call number. A subject index may also be added as there are times when it will be useful to know what journals are received by subject field. Call numbers for periodicals may also be obtained from the Author-Title section of the library's Main Card Catalog, or from alphabetical lists of titles which the library has on current subscription.

Next, consult and make brief annotations of what appear to be the most recent major articles that fully discuss the subject. Review articles, which survey the earlier literature, usually contain comprehensive bibliographies or footnotes of source items that may expedite the search, can be used to enlarge the scope of the search and show the historic development of the subject. English language abstracts of major foreign language articles should be noted and included so that full translations can be obtained. After reading the major articles, make a list of all the books and periodical articles from the bibliographies found in the articles.

Students searching the library frequently need to establish the full bibliographic description of a publication. A question results when presumed evidence for the existence of a publication is fragmentary, incomplete, unclear, or inconsistent. Complete, accurate citations are essential in obtaining copies of books or articles, and they are absolutely essential when citing references. Language, publication date, subject, place of publication can be verified in subject bibliographies, bibliographies, union catalogs or lists of serials, but subject-related indexes and abstracting services ought to be consulted first. Then consult each of the listed articles and their bibliographies until you find a repetition of the articles listed. Your search is ended, assuming the major article having the oldest date has been located. This type of library searching has been called a fan search because each major article contains a bibliography that spreads like a fan. The periodical articles from the first bibliography also list bibliographies that in turn fan out further until they all interconnect. Fan searching of articles is suggested when current information about your subject is not included in the periodical indexes, abstracts, and reviewing journals.

Brief news items, announcements of the marketing of a new product, preliminary announcements of developments of one kind or another, are the types of things that may not be considered worthy of being indexed or abstracted, yet they may have a great importance in certain kinds of searches. Scan the current issues of all periodicals for news departments and items that may pertain to your subject or an article written from a broader aspect that includes the specific subject, then proceed with a fan search.

Periodicals

Scientific and technological periodicals include: 1) society transactions, proceedings, journals and bulletins; 2) institutional journals, bulletins and newsletters; 3) technical journals; 4) trade journals; and 5) some house organs published by private corporations. All this variety of periodical literature may be found listed in a library's Serial Record. Articles found in these periodicals usually cover one subject in depth, describing methods, equipment and results more accurately than similar information found in reference books. The reason is that the author of a periodical article is usually the person who conducted the original research.

Periodicals contain the most recent available information regarding the current results of experimental research and announcements of scientific and technical developments. Current issues also present the latest market prices, biographical data, advertisements and small news items which are seldom covered by index and abstracting publications. Generally, it is years before the information contained in scholarly journals is rewritten into books, manuals and encyclopedias. Most important, the footnotes and bibliographies listed in recent articles serve as a guide to older information on the same subject.

Some back runs of periodicals are available in microform. This form of publication is a satisfactory way of accumulating back files because it saves storage space, and microform reader-printers can produce legible reproductions of the printed page. The availability of microform copies of periodicals may be learned from the announcements and catalogs of Microcard Editions, Inc., 901 Twenty-Sixth St., N.W., Washington, D.C. 20037. Its catalogs, Guide to Microforms in Print and Subject Guide to Microforms in Print, list most publications available in microform from commercial publishers in the U.S.A.

There may be difficulty in locating issues of journals published during the years of World War II (1939-1945). Photo-offset copies were made of some titles by J. W. Edwards of Ann Arbor, Michigan. However, all issues that were not readily available could not be duplicated for this project because of too limited demand for some titles. In an article in Research for April 1948 Cooper gave this information: "Complete runs of German periodicals covering the war period are to be found in Dutch, Swedish, and Swiss libraries, and photocopies may be purchased through the reproduction centers in the Hague (Federation Internationale de Documentation), Stockholm (Tekniska Hogskope), and Zurich (Eidgenossische Technische Hochschule)." It is likely that there will be at least one copy of all but the most obscure publications available in the United States. The Library of Congress makes every possible effort to procure them.

Ideally, libraries should have available all the periodicals in which scientific and technological articles are written. Unfortunately, the rapid rate in the growth of scientific publishing makes this ideal difficult, if not financially impossible, to accomplish. However, consultation of the following periodicals directories is suggested whenever new topics are being searched.

Brown, Peter & Stratton, George B. World list of scientific periodicals published in the years 1900-1960. Hamden, Conn., Shoe String, (Butterworth), 1964. 3 v.

New serials titles, a union list of serials commencing publication after December 31, 1949. Washington, D. C. U. S. Library of Congress Card Division, Navy Yard Annex. 1961-65 cumulation published by R. R. Bowker, New York.

Ulrich's international periodicals directory. 14th ed. New York, R. R. Bowker, 1970. 2 v.

Some house organs, the periodicals published by industrial firms, are available in academic libraries whose reference sources are as complete as possible. Many may be obtained by requesting that the name of the library be added to the mailing list; others are issued for limited distribution only, and occasionally there is a regular subscription charge. For detailed information, consult:

Market data book, annual issue of Industrial Marketing, 100 East Ohio Street, Chicago, Illinois.

The Printers' ink directory of house organs (1954). Printers' Ink Publishing Company, Inc., 205 East 42nd Street, New York.

The following list of scientific and technical periodicals, representative of the thousands of titles published each month, are covered by the leading indexing and abstracting services.

GENERAL SCIENCE

A. A. A. S. Bulletin. (American Association for Advancement of Science) 1942-1946. 1961--

American Academy of Arts and Sciences. Bulletin. v. 1, 1948--

American Journal of Science. v. 1, 1818--

California Academy of Sciences. Proceedings. v. 1, 1854--

Chicago Academy of Sciences. Bulletin. v. 1, 1883--

Daedalus. v. 1, 1955--

Endeavor. v. 1, 1942--

F. A. S. Newsletter. (Federation of American Scientists) v. 19, 1966--

International Science and Technology. v. 1, 1962--

Isis. v. 1, 1913--

Journal of Research in Science Teaching. v. 1, 1963--

Journal of Scientific Instruments. v. 1, 1923--

National Academy of Sciences. National Research Council. News Report. v. 1, 1951--

National Academy of Sciences. Proceedings. v. 1, 1915--

National Research Council. News Reports. v. 1, 1950--

New Research Centers. v. 1, 1965--

New York Academy of Sciences. Annals. v. 1, 1877--

__________. Transactions. v. 1, 1881--

Review of Scientific Instruments. v. 1, 1930--

Science. v. 1, 1880--

Science and Children. v. 1, 1963--

Science and Citizen. v. 1, 1958--

Science and Math Weekly. v. 1, 1960--

Science Digest. v. 1, 1937--

Science Education. v. 1, 1916--

Science Education News. v. 1, 1959--

Science News. v. 1, 1921--

Science Teacher. v. 1, 1934--

Science World. v. 1, 1957--

Sciences. v. 1, 1961--

Scientific American. v. 1, 1845--

Scientific Information Notes. v. 1, 1959--

Scientific Meetings. v. 1, 1956--

Times Science Review. v. 1, 1951--

UNESCO Courier. v. 1, 1948--

Understanding. v. 1, 1962--

Washington Academy of Sciences. Journal. v. 1, 1911--

MATHEMATICS

American Journal of Mathematics. v. 1, 1878--

The American Mathematical Monthly. v. 1, 1894--

American Mathematical Society. Bulletin. v. 1, 1894--

__________. Proceedings. v. 1, 1950--

__________. Transactions. v. 1, 1962--

American Statistical Association. Journal. v. 1, 1868--

Annals of Mathematical Statistics. v. 1, 1929--

Annals of Mathematics. Series 2. v. 1, 1928--

Association for Computing Machinery. Communications. v. 1, 1958--

__________. Journal. v. 1, 1954--

Chinese Mathematics Acta. v. 1, 1960--

Communications on Pure and Applied Mathematics. v. 1, 1948--

Duke Mathematical Journal. v. 1, 1935--

International Journal of Computer Mathematics. v. 1, 1964--

Journal of Applied Mathematics and Mechanics. Translation. v. 22, 1958--

Journal of Mathematical Analysis and Application. v. 1, 1960--

Journal of Mathematics and Physics. v. 1, 1921--

Journal of Recreational Mathematics. v. 1, 1968--

Mathematical Reviews. v. 1, 1940--

Mathematics Magazine. v. 1, 1926--

Mathematics of Computation. v. 1, 1943--

The Mathematics Teacher. v. 1, 1908--

Pacific Journal of Mathematics. v. 1, 1951--

Quarterly of Applied Mathematics. v. 1, 1943--

School Science and Mathematics. v. 1, 1901--

Scripta Mathematica. v. 1, 1933--

S.I.A.M. Review. (Society for Industrial and Applied Mathematics) v. 1, 1959--

Society for Industrial and Applied Mathematics. Journal. v. 1, 1953--

__________. Journal. Series A: Control. v. 1, 1963--

Soviet Mathematics - Doklady. Translation. v. 1, 1960--

Theory of Probability and its Applications. Translation. v. 1, 1956--

U.S. National Bureau of Standards. Journal of Research: B: Mathematics and Mathematical Physics. v. 65, 1961--

ASTRONOMY

The Astronomical Journal. v. 1, 1849--

The Astrophysical Journal. v. 1, 1895--

Comments on Astrophysics and Space Physics. v. 1, 1969--

Planetarium. v. 1, 1968--

Review of Popular Astronomy. v. 1, 1951--

Sky and Telescope. v. 1, 1941--

Soviet Astronomy. Translation. v. 1, 1957--

Space World. v. 1, 1962--

Vistas in Astronomy. v. 1, 1955--

PHYSICS

Academy of Sciences, U.S.S.R. Physical Series. Bulletin. v. 1, 1954--

Acoustical Society of America. Journal. v. 1, 1929--

American Journal of Physics. v. 1, 1933--

Applied Optics. v. 1, 1962--

Applied Physics Letters. v. 1, 1962--

Comments on Nuclear and Particle Physics. v. 1, 1967--

Comments on Solid State Physics. v. 1, 1968--

Contemporary Physics. v. 1, 1959--

Infrared Physics. v. 1, 1961--

Journal of Applied Physics. v. 1, 1931--

Journal of Mathematical Physics. v. 1, 1960--

Journal of Physics. v. 1, 1968--

Journal of the Mechanics and Physics of Solids. v. 1, 1952--

Optical Society of America. Journal. v. 1, 1917--

Physical Review. v. 1, 1893--

Physics of Fluids. v. 1, 1958--

Physics of Metals and Metallography. v. 1, 1958--

Physics Teacher. v. 1, 1963--

Physics Today. v. 1, 1948--

Review of Scientific Instruments. v. 1, 1930--

Reviews of Modern Physics. v. 1, 1929--

Sound, Its Uses and Control. v. 1, 1962--

U.S. National Bureau of Standards. Journal of Research:
A. Physics and Chemistry. v. 65, 1961--
C. Engineering and Instrumentation. v. 65, 1961--
D. Radio Science. v. 65, 1961--

Vacuum. v. 1, 1951--

CHEMISTRY

Academy of Sciences, U.S.S.R. Chemistry Section. Proceedings. v. 1, 1956--

American Chemical Society. Journal. v. 1, 1879--

Analytical Biochemistry. v. 1, 1960--

Analytical Chemistry. v. 1, 1929--

Applied Spectroscopy Reviews. v. 1, 1967--

Biochemical and Biophysical Research Communications. v. 1, 1959--

Biochemistry. v. 1, 1962--

Chemical & Engineering News. v. 1, 1923--

Chemical Reviews. v. 1, 1924--

Chemistry. v. 1, 1927--

Education in Chemistry. v. 1, 1964--

Industrial and Engineering Chemistry. v. 1, 1909--

Inorganic Chemistry. v. 1, 1962--

Journal of Applied Chemistry. v. 1, 1950--

Journal of Biological Chemistry. v. 1, 1905--

Journal of Catalysis. v. 1, 1962--

Journal of Chemical and Engineering Data. v. 1, 1956--

Journal of Chemical Documentation. v. 1, 1961--

Journal of Chemical Education. v. 1, 1924--

Journal of Chemical Physics. v. 1, 1933--

Journal of Colloid and Interface Science. v. 1, 1946--

Journal of General Chemistry. Translation. v. 1, 1949--

Journal of Lipid Research. v. 1, 1959--

Journal of Molecular Spectroscopy. v. 1, 1957--

Journal of Organic Chemistry. v. 1, 1935--

Journal of Physical Chemistry. v. 1, 1896--

Journal of Polymer Sciences. v. 1, 1945--

Journal of Ultrastructure Research. v. 1, 1957--

Kinetics and Catalysis. Translation. v. 1, 1960--

Soviet Radiochemistry. Translation. v. 4, 1962--

EARTH SCIENCES

American Association of Petroleum Geologists. Bulletin. v. 1, 1917--

American Geophysical Union. Transactions. v. 1, 1919--

American Mineralogist. v. 1, 1916--

Earth Science. v. 1, 1946--

Economic Geology. v. 1, 1905--

Geological Society of America. Bulletin. v. 1, 1888--

Geomagnetism and Aeronomy. v. 1, 1961--

Geophysics. v. 1, 1936--

GeoScience Information Society. Newsletter. v. 1, 1967--

GeoTimes. v. 1, 1956--

Journal of Applied Meteorology. v. 1, 1952--

Journal of Atmospheric and Terrestrial Physics. v. 1, 1950--

Journal of Geological Education. v. 1, 1951--

Journal of Geology. v. 1, 1893--

Journal of Geophysical Research. v. 1, 1896--

Oceanography, the Weekly of the Ocean. v. 1, 1963--

Soviet-Bloc Research in Geophysics, Astronomy and Space. v. 1, 1961--

Weatherwise; a Magazine about Weather. v. 1, 1948--

BIOLOGICAL SCIENCES

Academy of Natural Sciences, Philadelphia. Proceedings. v. 116, 1964--

American Biology Teacher. v. 1, 1938--

American Journal of Botany. v. 1, 1914--

American Museum of Natural History. Bulletin. v. 1, 1881--

American Naturalist. v. 1, 1867--

Audubon Magazine. v. 1, 1899--

Bioastronautics Report. v. 1, 1962--

Biophysical Journal. v. 1, 1960--

Bioscience. v. 1, 1951--

Ecology. v. 1, 1920--

Federation of American Societies for Experimental Biology. Federation Proceedings. v. 1, 1958--

Human Biology. v. 1, 1929--

Journal of Animal Ecology. v. 1, 1920--

Journal of the History of Biology. v. 1, 1968--

National Geographic. v. 1, 1899--

Natural History. (Incorporating: Nature Magazine.) v. 1, 1900--

Nature and Science. v. 1, 1963--

Society for Experimental Biology and Medicine. Proceedings. v. 1, 1903--

Stain Technology. v. 1, 1925--

Vision Research. v. 1, 1961--

Yale Journal of Biology and Medicine. v. 1, 1928--

MEDICAL SCIENCES

Aerospace Medicine. v. 1, 1930--

American Dental Association. Journal. v. 1, 1913--

American Journal of the Medical Sciences. v. 1, 1820--

American Medical Association. Journal. v. 1, 1883--

Annals of Internal Medicine. v. 1, 1922--

Hospital Topics. v. 1, 1922--

Journal of Medical Education. v. 1, 1926--

Lancet. v. 1, 1823--

Medical Electronics & Biological Engineering. v. 1, 1963--

Medical Library Association. Bulletin. v. 1, 1911--

Medical Technology. v. 1, 1967--

New England Journal of Medicine. v. 1, 1812--

Today's Health. v. 1, 1923--

AGRICULTURAL SCIENCES

Agricultural Education Magazine. v. 1, 1929--

Agricultural History. v. 1, 1927--

American Dairy Review. v. 1, 1939--

American Dietetics Association. Journal. v. 1, 1925--

American Forests. v. 1, 1895--

American Journal of Agricultural Economics. v. 1, 1919--

American Veterinary Medical Association. Journal. v. 1, 1869--

Association of Official Agricultural Chemists. Journal. v. 1, 1915--

Food Technology. v. 1, 1947--

The Garden Journal. v. 1, 1951--

International Association of Agricultural Librarians and Documentalists. Quarterly Bulletin. v. 1, 1956--

Journal of Dairy Science. v. 1, 1936-- (Contains Abstracts of Literature on Milk and Milk Products.)

Journal of Forestry. v. 1, 1902--

Journal of Home Economics. v. 1, 1909--

Journal of the Science of Food and Agriculture. v. 1, 1950-- (Abstracts cover chemistry of agriculture, food and sanitation.)

Nutrition Reviews. v. 1, 1942--

Pesticides Monitoring Journal. v. 1, 1967--

Prairie Farmer. v. 1, 1841--

Soil Science. v. 1, 1916--

ENGINEERING SCIENCES

AIAA Bulletin. (American Institute of Aeronautics and Astronautics) v. 1, 1964--

AIAA Journal. (American Institute of Aeronautics and Astronautics) v. 1, 1963--

AIChE Journal. (American Institute of Chemical Engineers) v. 1, 1955--

ASHRAE Journal. (American Society of Heating, Refrigeration and Air Conditioning Engineers) v. 1, 1914--

Aerospace. v. 1, 1963--

AeroSpace Engineering. v. 1, 1942--

Air Conditioning, Heating and Ventilating. v. 1, 1904--

Air Pollution Control Association. Journal. v. 1, 1951--

American Association of Petroleum Geologists. Bulletin. v. 1, 1917--

American Aviation. v. 1, 1937--

American Builder. v. 1, 1879--

American Ceramic Society. Bulletin. v. 1, 1818--

__________. Journal. v. 1, 1918--

American Concrete Institute. Journal. v. 1, 1929--

American Dyestuff Reporter. v. 1, 1917--

American Gas Association Monthly. v. 1, 1912--

American Machinist. v. 1, 1877--

American Society of Civil Engineers. Proceedings. v. 1, 1873--
Consists of Journals of following divisions:
- Aero-Space-Transport
- City Planning (superseded by Urban Planning and Development)

Construction
Engineering Mechanics
Highway
Hydraulics
Irrigation and Drainage
Pipeline
Power
Sanitary Engineering
Soil Mechanics and Foundations
Structural
Surveying and Mapping
Waterways and Harbors
Professional Practice

American Society of Naval Engineers. Journal. v. 1, 1889--

American Society of Safety Engineers. Journal. v. 1, 1956--

American Water Works Association. Journal. v. 1, 1914--

Analytical Chemistry. v. 1, 1929--

Applied Mechanics Reviews. v. 1, 1948--

Astronautics & Aeronautics. v. 1, 1957--

Atmospheric Environment. v. 1, 1958--

Audio. v. 1, 1917--

Audio Engineering Society. Journal. v. 1, 1953--

Automatic Control. v. 1, 1954--

Automation. v. 1, 1954--

Automation and Remote Control. v. 25, 1964--

Aviation Week & Space Technology. v. 1, 1916--

Bell Laboratories Record. v. 1, 1925--

Bell System Technical Journal. v. 1, 1922--

Bulletin of the Atomic Scientists. v. 1, 1945--

Ceramic Industry. v. 1, 1923--

Chemical & Engineering News. v. 1, 1923--

Chemical Engineering. v. 1, 1902--

Chemical Engineering Progress. v. 1, 1947--

Chemical Industries. v. 1, 1914--

Chemical Week. v. 1, 1914--

Civil Engineering. v. 1, 1930--

Coal Age. v. 1, 1911--

Combustion. v. 1, 1929--

Computer Design. v. 1, 1962--

Computers and Automation. v. 1, 1951--

Construction Methods and Equipment. v. 1, 1919--

Control Engineering. v. 1, 1954--

Corrosion. v. 1, 1945--

Data Systems Design. v. 1, 1963--

Design News. v. 1, 1946--

Electrical Engineering. v. 1, 1887--

Electrical World. v. 1, 1874--

Electrochemical Society. Journal. v. 1, 1948--

Electrochemical Technology. v. 1, 1963--

Electronics. v. 1, 1930--

Electronics World. v. 1, 1919--

Electro-Technology. v. 1, 1928--

Engineer. v. 1, 1956--

Engineering and Mining Journal. v. 1, 1866--

Engineering Journal. v. 1, 1964--

Engineering News-Record. v. 1, 1874--

Food Engineering. v. 1, 1928--

Franklin Institute. Journal. v. 1, 1826--

Heating, Piping & Air Conditioning. v. 1, 1929--

Hydraulics and Pneumatics. v. 1, 1948--

IEEE Proceedings. v. 1, 1913--

IEEE Spectrum. v. 1, 1964--

IEEE Transactions. (Formerly: Institute of Radio Engineers. Transactions.) v. 1, 1952--
In 35 groups:
Aero-Space
Aerospace and Navigational Electronics
Antennas and Propagation
Audio
Automatic Control
Bio-Medical Engineering
Broadcast and Television Receivers
Broadcasting
Circuit Theory
Communication Technology
Component Parts
Computer
Education
Electrical Insulation
Electromagnetic Compatibility
Electron Devices
Engineering Management
Engineering Writing and Speech
Geoscience Electronics
Human Factors in Electronics
Industrial Electronics and Control Instrumentation
Industry and General Applications
Information Theory
Instrumentation and Measurement
Magnetics
Microwave Theory and Techniques

Military Electronics
Nuclear Science
Power
Product Engineering & Production
Reliability
Sonics and Ultrasonics
Space Electronics and Telemetry
Systems Science & Cybernetics
Vehicular Communications

Industrial and Engineering Chemistry. v. 1, 1909--

Industrial and Engineering Chemistry Fundamentals. v. 1, 1962--

Industrial Quality Control. v. 1, 1944--

Instrumentation Technology. v. 1, 1954--

Instruments and Automation. v. 1, 1928--

Instruments and Control Systems. v. 1, 1928--

I. S. A. Journal. (Instruments Society of America) v. 1, 1954--

International Chemical Engineering. v. 1, 1961--

Iron Age. v. 1, 1855--

Journal of Aerospace Sciences. v. 1, 1934--

Journal of Applied Mechanics. v. 1, 1935--

Journal of Basic Engineering. v. 81, 1959--

Journal of Biochemical and Microbiological Technology and Engineering. v. 1, 1959--

Journal of Electronics. v. 1, 1935--

Journal of Engineering Education. v. 1, 1910--

Journal of Engineering for Industry. v. 81, 1959--

Journal of Engineering for Power. v. 81, 1959--

Journal of Heat Transfer. v 81, 1959--

Journal of Industrial Engineering. v. 1, 1949--

Journal of Metals. v. 1, 1949--

Journal of Petroleum Technology. v. 1, 1948--

Journal of Spacecraft and Rockets. v. 1, 1964--

Journal of Terramechanics. v. 1, 1964--

Magazine of Standards. v. 1, 1930--

Materials in Design Engineering. v. 1, 1929--

Materials Research & Standards. v. 1, 1961--

Mechanical Engineering. v. 1, 1906--

Mechanics Illustrated. v. 1, 1928--

Metal Progress. v. 1, 1920--

Microwave Journal. v. 1, 1958--

Mining Engineering. v. 1, 1949--

Missiles and Rockets. v. 1, 1956--

Modern Plastics. v. 1, 1925--

Nucleonics. v. 1, 1947--

Popular Electronics. v. 1, 1954--

Popular Mechanics Magazine. v. 1, 1902--

Popular Science Monthly. v. 1, 1872--

Power. v. 1, 1882--

Power Engineering. v. 1, 1896--

Product Engineering. v. 1, 1930--

Public Roads. v. 1, 1918--

QST. v. 1, 1915--

RCA Review. v. 1, 1936--

Radio and Television Weekly. v. 1, 1916--

Radio Engineering. Translation. v. 1, 1957--

Radio Engineering and Electronics. Translation. v. 1, 1957--

Railway Age. v. 1, 1856--

Research/Development Magazine. v. 1, 1950--

Research Management. v. 1, 1958--

Roads and Streets. v. 1, 1906--

S. A. E. Journal. (Society of Automotive Engineers) v. 1, 1917--

S. M. P. T. E. Journal. (Society of Motion Pictures and Television Engineers) v. 1, 1916--

S. P. E. Journal. (Society of Plastics Engineers) v. 1, 1945--

Society of Petroleum Engineers. Journal. v. 1, 1961--

Solid-State Electronics. v. 1, 1960--

Space/Aeronautics. v. 1, 1943--

Steel. v. 1, 1882--

Telecommunications and Radio Engineering. Translation. v. 1, 1962--

Textile Research Journal. v. 1, 1930--

Tool and Manufacturing Engineer. v. 1, 1932--

Water Pollution Control Federation. Journal. v. 1, 1928--

Water Works & Waste Engineering. v. 1, 1944--

Indexes to Dissertations and Theses from Academic Institutions

For many years the information buried in doctoral dissertations was almost inaccessible to the library searcher. Thousands of dissertations were carefully prepared, submitted, published and then lost in archives and libraries around the world. Each year doctoral candidates in leading colleges and universities contribute significantly to the extension of man's knowledge through the requirements for doctoral dissertations. Although a few institutions still require ink-print publication for dissertations, almost all colleges and universities granting doctoral degrees now have given up this requirement in favor of microfilm publication. Since 1938, this great wealth of useful knowledge has been indexed and published by University Microfilms on microfilm and in Dissertation Abstracts International, listing 30,000 new abstracts yearly. The development of xerography in the 1950's made it possible to obtain dissertation copies on paper or microfilm. Compiling a bibliography of relevant dissertations formerly meant carefully searching through each of the volumes of Dissertation Abstracts. Such a search required days and weeks of searching for the most perceptive researchers and the librarians assisting them.

Doctoral degree candidates and library searchers in all fields can now put to work for themselves an information retrieval system, DATRIX ("Direct Access to Reference Information; A Xerox Service"), developed by University Microfilms. DATRIX takes stock of the entire Dissertation Abstracts International file and pinpoints any dissertation or group of dissertations relevant to a particular subject.

DATRIX is simple in operation even though its data base encompasses thousands of doctoral dissertations, comprising the majority of all dissertations published since 1938. (Plans call for expansion of the data base to include all American and Canadian dissertations published since 1861.) To enable it to hunt most effectively and economically throughout the expansive information area, the DATRIX computer's memory is filled with a constant input of key words derived from the dissertation titles, author's selected subject headings and other descriptive data.

Dissertations are classified into two broad areas: "Sciences/Engineering" and "Humanities/Social Sciences." A single comprehensive key word list covering both areas provides the foundations from which questions are quickly formulated for the computer. With the publication of the Retrospective Index, Volumes I-XXIX to dissertations accepted in partial fulfillment of doctoral degrees in American universities, the user has a single source index to the major portion of this vast body of scholarly information.

In the Retrospective Index the principal words in each dissertation title are printed and used as keywords in alphabetical order in columns placed under subject categories. No restriction can be placed on the words that writers of dissertations use in their titles, so no restriction can be placed on the words which appear as keywords in this index. Many common words not relevant to retrieval have been omitted from the list: articles, connectives, and prepositions, such as a, an, the, of, for, etc.

To make this index easier to use, the complete Retrospective Index has been divided into broad subjects arranged as follows:

Volume I	Mathematics & Physics
Volume II	Chemistry
Volume III	Earth/Life Sciences
Volume IV	Psychology/Sociology/Political Science
Volume V	Social Sciences
Volume VI	Engineering
Volume VII	Education
Volume VIII	Communication/Information/Business Literature/Fine Arts
Volume IX	Author Index

Once the more promising dissertation titles and citations from this index are selected, turn to volume-issue and page in Dissertation Abstracts International (DAI) to read the dissertation abstract. This index provides the complete volume and page reference to DAI, or to Dissertation Abstracts or Microfilm Abstracts, which are older titles of DAI.

Dissertation Abstracts International. (Formerly Dissertations Abstracts) Section A. Humanities and Social Sciences; Section B. Physical Sciences and Technology. Ann Arbor, Mich., University Microfilms, Xerox Co., 1938--

v. 1-11 (1938-51) Microfilm Abstracts
1952-July 1969 Dissertation Abstracts
1956-1964/65 a 13th number was issued, called the Index to American Doctoral Dissertations. (1963/64-1964/65 American Doctoral Dissertations) This includes the index to Dissertation Abstracts for the year and follows Doctoral Dissertations accepted by American Universities.
v. 27 issued in 2 sections as above.

____ Microfilm Abstracts Author Index, v. 1-11, 1938-1951. comp. by the Georgia Chapter, Special Libraries Assn. with the cooperation of University Microfilms, Ann Arbor, Mich., Atlanta, 1956.

____ Retrospective Index, Volumes I-XXIX. Ann Arbor, Mich., University Microfilms, c1970-- 9 v.

The listing should be searched by any graduate student defining a dissertation subject, to be sure that it has not been covered before from the same viewpoint and that a previous study has not disproved his theory. Such information not only adds to his scholarly resources, but also helps prevent duplicated effort.

American Chemical Society. Directory of graduate research: faculties, publications, and doctoral theses in departments of chemistry, biochemistry, and chemical engineering at United States universities. Washington, D. C., 1957--

Doctoral dissertations accepted by American universities, 1933-1957. New York, H. W. Wilson Co., 1957.

Marckworth, M. Lois, comp. Dissertations in physics, an indexed bibliography of all doctoral theses accepted by American universities, 1861-1959. Stanford, Cal., Stanford Univ. Press, 1961.

U. S. Library of Congress. List of American doctoral dissertations printed in 1912-1938. Washington, D. C., U. S. G. P. O., 1913-1940. 27 v.

________. National union catalog of manuscript collections; based on reports from American repositories of manuscripts. Ann Arbor, Mich., J. W. Edwards, 1959-1961--

Other guides and bibliographies of theses and manuscripts are also available for research:

Black, Dorothy M., comp. Guide to lists of masters theses. Chicago, American Library Association, 1965.

Masters Abstracts; catalog of selected master theses on microfilm. Ann Arbor, Mich., University Microfilms, Xerox Co., 1962--

The National Archives. Guides to German records microfilmed at Alexandria, Va. Washington, D. C., U. S. G. P. O., 1958-1968-- (GS 4. 2:631/1-59--)

__________. List of record groups in the National Archives and the Federal Records Centers. Washington, D. C., U. S. G. P. O., February 1969 (GS4. 19:969)

Office of the National Archives. The Administration of modern archives: a select bibliographic guide. Washington, D. C., U. S. G. P. O., 1970.

U. S. Library of Congress. National register of microcopy masters. Washington, D. C., U. S. G. P. O., 1965--

Government Documents

The term "government document" or "government publication" covers reports, bulletins and publications issued by a municipal, state, or federal agency. Academic libraries usually maintain a document "depository" collection published by U.S. federal agencies. These "depository" collections had their beginnings in the 1859 act that charged the Department of the Interior with the publication and distribution of government documents. In general, a depository library is one that receives a copy of every government publication issued, with certain exceptions. These exceptions include documents classified as secret, confidential, or for restricted circulation; much "processed" material, not printed but reproduced by mimeography or similar process; considerable graphic material, such as maps, posters, photographs, films, and the like; various ephemera; and parts of publications that appear in separate issues but are intended for binding later in single units.

The U.S. Government Organization Manual (Washington, D.C., U.S.G.P.O., 1935--) contains an appendix listing more than 150 important periodicals published by different departments and agencies of the Federal government on a regular basis.

How to Find Government Publications

First check the Author-Title section of the library's Main Card Catalog. Many government publications are catalogued under the name of the country or state, followed by the name of the agency. Examples (The titles are underlined):

Canada. Department of Transport. Meteorological Division. Meteorological summary for Dorval, Quebec.

Great Britain. Parliament. The Parliamentary debates.

Louisiana. Department of Conservation. Wildlife resources of Louisiana; their nature, value and protection.

U. N. Economic Commission for Europe. Competition between steel and aluminum.

U. S. Public Health Service. Division of Special Health Services. Bibliography of occupational health.

If the U. S. document you are seeking is not listed in the library's Main Card Catalog, consult the Monthly Catalog of United States Government Publications (Washington, D. C., U. S. G. P. O., 1895--) which is the major index to United States government publications. Each month's issue has an index which is cumulated annually. The catalog lists over 20,000 separate titles published each year, plus nearly every government publication issued by the U. S. G. P. O. since 1895. To find U. S. Government documents on any subject published by the major long-established federal agencies, search and record the document numbers as they are located under the name of the issuing agency.

Price Lists of Government Publications, each covering a different subject and listing government publications about the subject regardless of the issuing agency, are available from the U. S. Superintendent of Documents, Washington, D. C. upon request. Although selective and limited in number, the lists are of value in calling attention to certain documents before the Monthly Catalog becomes available.

Titles of some of the Price Lists relating to science and technology are:

15 GEOLOGY
19 ARMY. Field manuals and technical manuals.
21 FISH AND WILDLIFE.
25 TRANSPORTATION, HIGHWAYS, ROADS, AND POSTAL SERVICE.
31 EDUCATION.
36 GOVERNMENT PERIODICALS AND SUBSCRIPTION SERVICE.
38 ANIMAL INDUSTRY. Farm animals, poultry and dairying.
41 INSECTS. Worms and insects harmful to man, animals, and plants.
42 IRRIGATION, DRAINAGE, WATER POWER.
43 FORESTRY. Managing and using forest and range land including timber and lumber, ranges and grazing, American woods.

44 PLANTS. Culture, grading, marketing, and storage of fruits, vegetables, grass, and grain.
46 SOILS AND FERTILIZERS. Soil survey, erosion, conservation.
48 WEATHER, ASTRONOMY, AND METEOROLOGY.
51 HEALTH AND HYGIENE. Drugs and sanitation.
51A DISEASES. Contagious and infectious diseases, sickness, and vital statistics.
53 MAPS. Engineering, surveying.
55 SMITHSONIAN INSTITUTION. National Museum and Indians.
58 MINES. Explosives, fuel, gasoline, gas, petroleum, minerals.
62 COMMERCE. Business, patents, trademarks, and foreign trade.
63 NAVY, MARINE CORPS, AND COAST GUARD.
64 SCIENTIFIC TESTS, STANDARDS. Mathematics, physics.
68 FARM MANAGEMENT. Foreign agriculture, rural electrification, agricultural marketing.
79 AIR FORCE. Aviation, Civil Aviation, Naval Aviation, Federal Aviation Administration.
79A SPACE. Missiles, the Moon, NASA, and Satellites, Space Education, Exploration, Research and Technology.
82 RADIO AND ELECTRICITY. Electronics, radar, and communications.
83 LIBRARY OF CONGRESS.
84 ATOMIC ENERGY AND CIVIL DEFENSE.
86 CONSUMER INFORMATION. Family finances, appliances, recreation, gardening, health and safety, food, house and home, child care, clothing and fabrics.

Guides to Government Documents

Reviews of government documents may be found in non-government periodicals such as The Booklist, Library Journal, Wilson Library Bulletin, and Special Libraries. Some publications issued either by government agencies or as the result of federally funded research are indexed by the standard indexing and abstracting services. For indexes and abstracts relating to special subjects, see those subjects. References to government publications may be found in many periodical indexes and abstracts including: Applied Science and Technology Index, Biological Abstracts, Biological and Agricultural Index, Chemical Abstracts, Education Index, Engineering Index, Occupational Index and Public Affairs Information Service.

Because many documents are not included in the Main Card Catalog, finding information in government publications usually requires a search through printed guides, catalogs and indexes. Bibliographies of government documents are normally placed in the Library of Congress "Z" classification for Bibliography. A major area of science writing can be found in government technical reports. These documents are the result of government-sponsored research which began during World War II, contracted primarily by the Air Force, the Atomic Energy Commission, the Department of Agriculture, the Department of Commerce, the Office of Naval Research and other government agencies.

Publications of selected states, other countries and international organizations may also be located in the Main Card Catalog. If not in the Card Catalog, look up the Monthly Checklist of State Publications (Washington, D. C., U. S. G. P. O., 1910--) which records all state publications received by the Library of Congress. Many departments of the Federal government such as the National Bureau of Standards, the U. S. Bureau of Mines, the U. S. Department of Agriculture, and the U. S. Geological Survey, issue their own catalogs and lists of publications. Students will find these lists helpful as subject bibliographies.

Government agencies in countries other than the United States issue publications that may contain information of importance to a particular research problem if they can be located. In some instances there are bureaus such as the Department of Scientific and Industrial Research in Great Britain in which the divisions correspond to similar United States Government scientific research bureaus. The United Kingdom Atomic Energy Research Authority is a source in recent years of documents like those issuing from the United States Atomic Energy Commission. In fact, some British documents are listed in A. E. C. abstracts and bibliographies.

Publications of the United Nations and its specialized agencies are listed in the United Nations Documents Index, issued monthly and including two separate annual cumulations: The Cumulative Checklist and The Cumulative Index (New York, United Nations Publications, v. 1-- 1950--). For documents prior to 1950 see the United Nations, Secretariat,

Department of Public Information, Library Services, Checklist of United Nations Documents (Lake Success, New York, 1949) for the period 1946-1949.

There are thousands of research projects carried out under grants or contracts by educational, industrial and professional organizations. Many technical or research and development reports have never been published as journal articles, consequenty they are not included in standard indexes and abstracts. In addition to those titles previously mentioned, the following are some of the more useful abstracts, bibliographies, catalogs, guides and indexes designed to be of aid in locating and using government publications.

GENERAL

Andriot, John L. Guide to popular U.S. government publications. Arlington, Va., Documents Index, 1961.

__________. Guide to U.S. government serials and periodicals. McLean, Va., Documents Index, 1962-66.

__________. Guide to U.S. government statistics. Arlington, Va., Documents Index, 1956--

Brock, Clifton. The Quiet crisis in government publications. College and Research Libraries, 26: 477-489, 1965.

Childs, James B. Government publications. Library Trends, 15: 378-397, 1967.

Hartford, Peter J. and Osborn, Jeanne. The Government publications course: a survey. Journal of Education for Librarianship, vol. 11, no. 3: 251-260, 1971.

Heinritz, Fred J. The Present state of the teaching of government publications in library schools. Library Trends, 15: 157-166, July 1966.

Jackson, Ellen. A Manual for the administration of the federal documents collection in libraries. Chicago, American Library Assn., 1966 reissue, 1955.

__________. Subject guide to major United States government publications: a partial list of non-GPO imprints. Chicago, American Library Assn., 1968.

Kling, Robert E., Jr. The Government printing office. New York, Praeger, c1970. (Praeger Library of U.S. Government Departments and Agencies No. 26.)

Leidy, W. Philip. Popular guide to government publications. 3rd ed. New York, Columbia Univ. Pr., 1968.

McReynolds, Floris. Microforms of U.S. government publications. Univ. of Ill. Library School Occasional Papers no. 69, 1963.

Mechanic, Sylvia. Annotated list of selected United States government publications available to depository libraries. New York, H. W. Wilson, 1971.

Oller, Anna Kathryn. Government publications. Drexel Library Quarterly, vol. 1, no. 4, 1965.

One hundred G.P.O. years, 1861-1961, history of United States printing. Washington, D.C., U.S.G.P.O., 1961. (GP1.2:G74/7/861-961.)

Schmeckebier, Laurence F. and Eastin, Roy B. Government publications & their use. rev. ed. Washington, D.C., Brookings Institution, 1969.

Shaw, Thomas Shuler, ed. Federal, state and local publications. Library Trends, vol. 15, no. 1, 1966.

U.S. Library of Congress. Serial Division. Popular names of U.S. government reports, a catalog. Compiled by Donald F. Wisdom and William P. Kilroy, Washington, D.C., U.S.G.P.O., 1966.

White, Alex Sandri, comp. Thirty million books in stock: everyone's guide to U.S. government publications. Allenhurst, N.J., Aurea, 1970.

Wood, Jennings, ed. United States government publications: a partial list of non-GPO imprints. Chicago, American Library Assn., 1964.

GENERAL SCIENCE

Clearinghouse announcements in science & technology. U.S. Dept. of Commerce, Springfield, Va., Clearinghouse, v. 1-- 1968--

Government-wide index to federal research and development reports. U. S. Dept. of Commerce, Springfield, Va., Clearinghouse, 1965--

Scientific and technical aerospace reports (S. T. A. R.). National Aeronautics and Space Administration, Scientific and Technical Information Division, Washington, D. C., U. S. G. P. O., 1963--

United States government research and development reports. U. S. Dept. of Commerce, N. B. S. Institute of Applied Technology, Springfield, Va., Clearinghouse, 1946--

PHYSICS

Nuclear science abstracts. Oak Ridge, Tenn., U. S. Atomic Energy Commission, Division of Technical Information Extension, v. 1-- 1947--

EARTH SCIENCES

Geochemical prospecting abstracts. Washington, D. C., United States Geological Survey, 1953--. (U. S. G. S. Bulletins 1000A, 1000G, 1098B.)

Geophysical abstracts. Washington, D. C., U. S. G. P. O., vol. 1-- 1929--

Selected water resources abstracts. (Office of Water Resources Research, U. S. Dept. of the Interior.) Springfield, Va., Clearinghouse, v. 1-- 1968--

MEDICAL SCIENCES

Arthritis and rheumatic diseases abstracts. Washington, D. C., U. S. G. P. O., v. 1-- 1964--

Diabetes literature index. (National Institute of Arthritis and Metabolic Diseases.) Washington, D. C. U. S. G. P. O., v. 1-- 1966--

Endocrinology index. Washington, D. C., U. S. G. P. O., v. 1-- 1968--

Epilepsy abstracts. Washington, D. C., U. S. G. P. O., v. 1-- 1967--

Gastroenterology abstracts and citations. (National Institute of Arthritis and Metabolic Diseases.) Washington, D. C., U. S. G. P. O., v. 1-- 1966--

Index medicus. Washington, D. C., U. S. G. P. O. (Ed. and pub. by National Library of Medicine, Bethesda, Md.) v. 1-- 1960-- (Medical Subject Headings is published as Part 2 of Jan. Index Medicus.)

Mental retardation abstracts. (American Assn. on Mental Deficiency.) Washington, D. C., U. S. G. P. O., v. 1-- 1964--

Psychopharmacology abstracts. Washington, D. C., U. S. G. P. O., v. 1-- 1961--

AGRICULTURAL SCIENCES

Commercial fisheries abstracts. Seattle, Wash., Bureau of Commercial Fisheries, Division of Publications, v. 1-- 1948--

Fertilizer abstracts. Muscle Shoals, Ala., National Fertilizer Development Center, Tennessee Valley Authority, Technical Library, v. 1-- 1968--

Pesticides documentation bulletin. Beltsville, Md., Pesticides Information Center, National Agricultural Library, U. S. Dept. of Agriculture, v. 1-- 1965--

Sport fishery abstracts. Washington, D. C., U. S. Fish and Wildlife Service, Dept. of the Interior, v. 1-- 1955--

Wildlife review. Laurel, Md., Patuxent Wildlife Research Center, v. 1-- 1935--

ENGINEERING SCIENCES

Abstracts of instructional materials in vocational and technical education. Columbus, Ohio, Ohio State Univ., Center for Vocational and Technical Education, ERIC Clearinghouse, v. 1-- 1967--

Air University library index to military periodicals. Maxwell Air Force Base, Alabama, Air University Library, v. 1-- 1949--

Air University Quarterly review. Maxwell Air Force Base, Alabama, v. 1-- 1947--

Environmental effects on materials and equipment abstracts. Washington, D. C., National Academy of Science, National Research Council, Prevention of Deterioration Center, v. 1-- 1961--

Fire research abstracts and reviews. Washington, D. C., National Academy of Sciences, Committee on Fire Research, v. 1-- 1959--

H R I S Abstracts. (Highway Research Information Service) Washington, D. C., Highway Research Board, National Academy of Sciences, v. 1-- 1968--

Highway research abstracts. Washington, D. C., Highway Research Board, v. 1-- 1931--

Public health engineering abstracts. Washington, D. C., U. S. G. P. O., v. 1-- 1921--

Reliability abstracts and technical reviews (R. A. T. R.) Washington, D. C., National Aeronautics and Space Administration, Reliability and Quality Assurance Office, v. 1-- 1961--

Technical abstract bulletin. Alexandria, Va., Defense Documentation Center, v. 1-- 1957--

PATENTS

Andrew, Lucy Brett. Practical patent procedure. 7th ed. Jamaica, N. Y., author c/o Bruhn Bros., 1970.

Houghton, Bernard. Technical information sources: a guide to patents, standards and technical reports in literature. 2nd rev. & enl. ed. Hamden, Conn., Shoe String (Linnet), 1971.

Jones, S. V. Inventor's patent handbook. New York, Dial, 1967. (Revision of "You Ought to Patent That," 1962.)

U.S. Department of Commerce. General information concerning patents. Washington, D.C., U.S.G.P.O., 1966.

U.S. Patent Office. Manual of classification of patents. Washington, D.C., U.S.G.P.O., 1947--

__________. Official gazette. Washington, D.C., U.S. G.P.O., v. 1-- 1872--

__________. Official gazette. Index of patents. Washington, D.C., U.S.G.P.O., v. 1-- 1920--

__________. Official gazette - patent abstract section. Washington, D.C., U.S.G.P.O., 1968--

Chapter V

BIBLIOGRAPHY FORMS

Subject Bibliographies

The result of a library search on a scientific topic is the compiling of a state-of-the-art or subject bibliography: a systematized list of references arranged according to an accepted style, identifying and describing the sources of information on the subject. Source lists can include not only books but published periodical and newspaper articles, pamphlets, government documents, reports, theses, dissertations, microforms, film strips and recordings, or unpublished manuscripts, papers and speeches. If full bibliographical information about each source was recorded during the library search, there is no need to document again such bibliographic details as edition, date of publication, author's full name or complete periodical title in a bibliographical description of authorities consulted. The information should be presented in such a manner that the reader can readily identify and eventually locate any book or library material listed as sources for further or more detailed study. It is helpful to locate reviews of the books listed in order to obtain an expert appraisal of the value of these sources and to learn of other aspects of the subject which the authors may have overlooked.

If annotations are not included in your bibliography, then give the abstract or index periodical citation so the person reading the report can find and read the abstract to determine the relevance of the periodical article.

Bibliographic Description

Usually sources are arranged alphabetically according to the surname of the author, if available; otherwise by the title of the pamphlet, document or report that has no author. The following bibliographical entries have all information in complete form:

1. BOOKS:

(1) The name of the author, last name first,
(2) the title of the book as it appears on the title page,
(3) the edition, if other than the first,
(4) the number of volumes in the set if the whole set is used; if a single book, volume is not given,
(5) the place of publication,
(6) the name of the publisher,
(7) the date of publication, and
(8) the pages consulted or total pages of the book.

A. Personal Authors

If the author has one given name, write it in full; if more than one, write his first name and then his initial or initials as listed on the title page or catalog card. This applies to all bibliographical references, whether for books, magazine articles, reference book articles, etc.

If a book has two or three authors, invert only the first author's name, for alphabetical purposes in the bibliography.

B. Corporate Authors

(1) The name of the author of the publication -- department, bureau, or organization,
(2) the name of the document or publication,
(3) the edition, if other than the first,
(4) the place of publication,
(5) the publisher--not abbreviated, and
(6) the date of publication.

Abbreviations:

comp.--compiler, compiled
c--copyright
ed.--edition, editor, edited
enl.--enlarged
il.--illustrated, illustrator
introd.--introduction
n.d.--no date
n.p.--no place of publication
p.--page
pp.--pages
pref.--preface
pseud.--pseudonym
rev.--revised
tr.--translated, translator
v.--volume
v.--volumes

2. ARTICLES FROM REFERENCE BOOKS:

(1) The name of the author of the article, if known,
(2) the name of the article as it appears in the book,
(3) the name of the book in which the article appears,
(4) the edition, if other than the first, or the date of publication, or the copyright date,
(5) the volume number, if one of a set of books, and
(6) the inclusive paging of the article.

3. ARTICLES FROM PERIODICALS AND SERIALS:

(1) The name of the author of the article, if known,
(2) the title of the article,
(3) the full title of the serial in which the article appears,
(4) the volume of the magazine,
(5) the date, and
(6) the inclusive paging of the article.

In conforming to the conventional style in acknowledging sources of information, scientists have adopted abbreviated periodical citations in their research papers. Fortunately, two widely used standard lists of abbreviated references are: "A List of Periodicals Abstracted" found in Chemical Abstracts, for the physical and technical sciences; and the World List of Scientific Periodicals, published between 1900 and 1960 for the biological and natural sciences.

When in doubt about author, title or alphabetical arrangement, consult the book's catalog card in the library's Main Card Catalog as your best source of information.

Manuals of Style

Manuals of style, handbooks of research, manuscript preparation manuals, periodical guides or standards for bibliographical form explain the elements of a bibliographical citation, give reasons for documenting sources of information, include bibliographic examples for various types of publications and show bibliographical periodical abbreviation practice in different subject areas.

GENERAL SCIENCE

Allred, Dorald M. & Berna B. Guide for writing & typing theses & dissertations. Provo, Utah, Brigham Young Univ. Pr., 1969.

American National Standards Institute. American National Standard for the abbreviation of titles of periodicals. New York, 1969.

Baker, Sheridan. The Practical stylist. 2nd ed. New York, T. Y. Crowell, 1969.

Campbell, William G. Form and style in thesis writing. 3rd ed. Boston, Houghton Mifflin, 1969.

Dugdale, Kathleen. Manual of form for theses & term reports. 4th ed. Bloomington, Ind., Dugdale, 1967.

Grewe, Eugene F. & Sullivan, John F. The College research paper. 4th ed. Dubuque, Iowa, Wm. C. Brown, 1957.

Guide for Wiley authors in the preparation of linear auto-instructional programs. 3rd ed. New York, Wiley & Sons, 1967.

Hilbish, Florence. The Research paper. New York, Twayne, 1952.

Hook, Lucyle & Gaver, Mary V. The Research paper: gathering material; organizing and preparing the manuscript. 4th ed. New York, Prentice-Hall, 1969.

Kinney, Mary R. The Abbreviated citation: a bibliographical problem. Chicago, American Library Assn., 1967. (ACRL Monograph No. 28.)

McGraw-Hill Book Company, ed. McGraw-Hill author's book. New York, McGraw-Hill, 1968.

Mandel, Siegfried. Writing for science & technology. New York, Delta Books, Dell Publishing Co., 1970.

Perrin, Porter G., et al. Writer's guide and index to English. 4th ed. New York, Scott Foresman, 1965.

Stokes, Roy. The Function of bibliography. London, British Book Centre, 1969.

Strunk, William S., Jr. and White, E. B. The Elements of style. New York, Macmillan, 1962. (Macmillan Paperbacks 107.)

Thomas, Payne E. Guide for authors: manuscript, proof and illustration. Springfield, Ill., C. C. Thomas, 1968.

Turabian, Kate L. Manual for the writers of term papers, theses & dissertations. 3rd ed. Chicago, Univ. of Chicago Pr., 1967.

__________. Student's guide for writing college papers. 2nd ed. rev. Chicago, Univ. of Chicago Pr., 1970.

U.S. Government printing office style manual. Washington, D.C., U.S.G.P.O., 1967.

University of Chicago Press Editors. Manual of style. 12th rev. ed. Chicago, Univ. of Chicago Pr., 1969.

Williams, Cecil B. & Stevenson, Allan H. Research manual: for college studies and papers. 3rd ed. New York, Harper & Row, 1951.

MATHEMATICS

The American Mathematical Monthly. "Manual for Monthly Authors." Ed. by Harley Flanders. Washington, D.C., The American Mathematical Monthly, v. 78, no. 1, (January 1971), pp. 1-10. (Copies available upon request from the Mathematical Association of America, 1225 Connecticut Avenue, N. W., Washington, D. C. 20036.)

American Mathematical Society. A Manual for authors of mathematical papers. rev. ed. Buffalo, SUNY, 1966.

Journal of Mathematics. "On the Preparation of Manuscripts." Published inside back cover of each issue of Journal of Mathematics.

Mathematical Reviews. Abbreviations of the names of scientific periodicals. Buffalo, American Mathematical Society, 1970 reprint.

The Preparation and typing of mathematical manuscripts. New York, Bell Telephone Labs, 1963.

ASTRONOMY

The Astronomical Journal. Published inside front cover of each issue of The Astronomical Journal. "Style should be that of the American Institute of Physics publications, AIP Style Manual..."

The Astrophysical Journal. Published inside front cover of each issue: "Articles intended for publication in The Astrophysical Journal and its supplements should be prepared in accordance with the recommendations set forth in a Manual of Style available from the Editorial Office."

PHYSICS

American Institute of Physics. Style manual for guidance in the preparation of papers for journals published by the American Institute of Physics. New York, 1951, pp. 1-28. See particularly pp. 5, 14-18. (NOTE: Periodical abbreviations revised.)

CHEMISTRY

Access. Washington, D. C., American Chemical Society, v. 1, 1969--

American Chemical Society. "A List of Periodicals Abstracted." *Chemical Abstracts,* XLV (December 25, 1951), i-cclv.

__________. Literature of chemical technology. Washington, D. C., 1969. (Advances in Chemistry Series No. 78.)

__________. "Notice to Authors of Papers re: References, Abbreviations, etc." *Journal*, LXXIII (February 15, 1951), ii.

__________. "Suggestions to Authors of Review Articles." *Chemical Reviews,* XLVIII (February, 1951), pp. 1-5.

American Society of Biological Chemists. "Instructions to Authors." Published in preliminary pages of each issue of the *Journal of Biological Chemistry.*

Journal of Organic Chemistry. "Notice to Authors." Published in preliminary pages of each issue of the *Journal of Organic Chemistry*.

Journal of Physical Chemistry. "Notes." Published inside front cover of each issue of the *Journal of Physical Chemistry.*

Mellon, M. G. Chemical publications; their nature and use. 4th ed. New York, McGraw-Hill, 1965.

EARTH SCIENCES

American Journal of Science. "Instructions to Contributors." Published inside front cover of each issue of *American Journal of Science*.

Journal of Geology. "Suggestion to Contributors." Published in preliminary pages of each issue of *Journal of Geology*.

U. S. Geological Survey. Suggestions to authors of papers submitted for publication by the United States Geological Survey with directions to typists. By George McLane Wood, 4th ed. rev. and enl. by Bernard H. Lane; Washington, U. S. G. P. O., 1935, 126 pp. Frequently cited in the field as Lane, 1935, and as Wood and Lane, 1935. See particularly pp. 15-30.

Warthin, A. S., Jr. and Weller, J. Marvin. "Style, Form and Procedure for the Journal of Paleontology," *Journal of Paleontology*, XXV (September, 1951), pp. 709-712.

BIOLOGICAL SCIENCES

Botanical Society of America. "Notes for Contributors to the American Journal of Botany." Published on the back cover of each issue of the *American Journal of Botany*.

Conference of Biological Editors. Committee on Form and Style. Style manual for biological journals. 2nd ed. Washington, D. C., American Institute of Biological Sciences, 1964.

Madison Botanical Congress (1893). "Rules for Citation." Prepared by the Madison Botanical Congress and Section G, A. A. A. S., Torrey Botanical Club, *Bulletin,* XXII (March 27, 1895), pp. 130-132.

Schwarten, Lazella, comp. "Index to American Botanical Literature." With the collaboration of the editors of the *Taxonomic Index.* Published in each issue of the Torrey Botanical Club, *Bulletin*.

Schwarten, Lazella and Rickett, H. W., comps. "Abbreviations of Periodicals Cited in the Index to American Botanical Literature." Torrey Botanical Club, *Bulletin,* LXXIV (July, 1947), pp. 348-356.

Society of American Bacteriologists. "Instructions to Authors." Published in preliminary pages of each issue of *Journal of Bacteriology.*

__________. "Note to Contributors." Published in each issue of *Bacteriological Reviews*.

Torrey Botanical Club. "Instructions for Contributors." Published on the back cover of each issue of its *Bulletin.*

Wistar Institute of Anatomy and Biology, Philadelphia. A Guide for authors: The Wistar Institute journals. Philadelphia, n. d. See particularly pp. 2-3, paragraphs 10-13 for information on literature cited.

__________. Style brief. Philadelphia, 1934.

MEDICAL SCIENCES

American Medical Association. "(Notes to) Contributors." Published in each issue of its _Journal._

American Medical Association. Scientific Publications Division. Advice to authors. 2nd ed. Chicago, A. M. A., 1968.

American Psychosomatic Society. "General Information." Published inside cover of each issue of _Psychosomatic Medicine._

Bredow, Miriam. Medical secretarial procedures. 5th ed. New York, McGraw-Hill, 1966.

Fulton, J. R. "The Principles of Bibliographical Citation: An Informal Discourse Addressed to Writers of Scientific Papers." Medical Library Association _Bulletin,_ XXII (April, 1934), pp. 183-197.

List of periodicals indexed in _Index Medicus._ Washington, D. C., U. S. National Library of Medicine, v. 1, 1963--

U. S. Armed Forces Medical Journal. "Notice to Contributors." Published in preliminary pages of each issue of _United States Armed Forces Medical Journal._

U. S. Armed Forces Medical Library. Index catalogue of the surgeon-general's office. United States Army. Authors and Subjects (Series 1-4.) Washington, D. C., U. S. G. P. O., 1880--

AGRICULTURAL SCIENCES

Iowa State College of Agriculture and Mechanic Arts, Ames, Graduate College. Manual on thesis writing. 3rd ed. Ames, 1951.

U. S. National Agricultural Library. Serial publications indexed in Bibliography of agriculture. Rev. ed., Washington, D. C., 1965.

__________. Division of Indexing and Documentation. Bibliography of agriculture. Washington, D. C., v. 1,

1942-- (Dec. a subject and author index.)

ENGINEERING SCIENCES

American Institute of Chemical Engineers. Guide to authors. 2nd ed., New York, 1949. See particularly p. 3 for literature citations.

American Institute of Electrical Engineers. Information for authors preparation of manuscripts and presentation of papers. New York, 1948. See particularly pp. 8-9 for reference to literature.

American Society for Testing and Materials. Coden for 22,544 titles of serials. 2nd supplement. Philadelphia, 1969.

__________. Manual for authors of ASTM papers. Philadelphia, 1952. See particularly pp. 17-18 for footnotes and references.

American Society of Mechanical Engineers. An ASME paper: its preparation, submission and publication, and presentation. New York, 1951. (ASME Manual MS-4.)

__________. "(Directions re) Bibliography." Published on back cover of each issue of its Transactions.

American Water Works Association. "A Style Manual for Journal Authors." Journal, XXXIX (May, 1947), pp. 437-488.

Ehrlich, Eugene H. & Murphy, Daniel. The Art of technical writing: a manual for scientists, engineers and students. New York, Apollo Editions, 1969.

Franklin Institute, Philadelphia. "Suggestions to Authors of Papers for the Journal of the Franklin Institute." Published in the back of some of the recent issues of its Journal.

Freedman, George. A Handbook for the technical and scientific secretary. New York, Barnes & Noble, 1967.

Illuminating Engineering. "A Guide to Authors: Information for Authors on Preparation of Papers for Publication." Illuminating Engineering, XLI (September, 1946), pp. 648-653.

Institute of the Aeronautical Sciences. "Suggestions for Contributors to the Publications of the Institute of the Aeronautical Sciences." Published on back cover of each issue of its *Journal.*

Institute of Radio Engineers. "Preparation and Publication of I. R. E. Papers." Prepared by Helen M. Stote, *I. R. E. Proceedings,* XXXIV (January, 1946), pp. 5W-9W. See particularly p. 6W for information on footnote references.

Laird, Eleanor. Engineering secretary's complete handbook. 2nd ed. New York, Prentice-Hall, 1967.

Turner, Rufus P. Technical writer's & editor's stylebook. Indianapolis, Ind., Sams, 1964.

Weisman, Herman M. Basic technical writing. 2nd ed. New York, Bobbs Merrill, 1968.

AUTHOR INDEX

TITLE INDEX